Telecommunication Networks for the Smart Grid

For a complete listing of titles in the
Artech House Power Engineering Series,
turn to the back of this book.

Telecommunication Networks for the Smart Grid

Alberto Sendin

Miguel A. Sanchez-Fornie

Iñigo Berganza

Javier Simon

Iker Urrutia

ARTECH
HOUSE

BOSTON | LONDON
artechhouse.com

Library of Congress Cataloging-in-Publication Data
A catalog record for this book is available from the U.S. Library of Congress.

British Library Cataloguing in Publication Data
A catalogue record for this book is available from the British Library.

Cover design by John Gomes

ISBN 13: 978-1-63081-046-7

© 2016 ARTECH HOUSE
685 Canton Street
Norwood, MA 02062

10 9 8 7 6 5 4 3 2 1

To our families

Contents

Preface

The Scope

The smart grid is not a novel concept; it has existed in the electric power industry for more than a decade now. However, the many aspects of smart grids and their complexity have resulted in few comprehensive models or real deployments. The various technologies that define smart grids have evolved in recent years, technically and economically, creating different approaches for solutions and architectures. Many utilities around the world have initiated a slow but decisive path into the adoption of smart grid applications in their systems.

Telecommunication is one of the main technologies used for the deployment of smart grids. The objective of this book is to help electric power industry stakeholders and practitioners to understand the various telecommunication technologies and services available in the context of smart grids. The fast-changing nature of smart grids and multiple requirements need a solid telecommunication architecture definition that supports grid applications and solution expansion.

The Scene

This book is designed to be helpful for professionals both in the electric power and telecommunication areas and it uses concepts from both domains.

From the electric power systems perspective, the *grid* is an electricity network of synchronized power producers and consumers connected by transmission and distribution lines. The *smart grid* utilizes information and telecommunication technologies, including distributed computing and associated sensors and actuators, to improve its operation. *Applications* (i.e., software running in central or distributed servers) offer the tools to realize this improvement.

The grid consists mainly of power lines and substations. Two main categories of substations are going to be considered. *Primary substations* are the substations on the border between transmission and distribution segments, which supply power to groups of *secondary substations*, the distribution network sites that ultimately provide the final transformation to the low voltage (LV) network, which supplies final customers.

The grid is operated by a number of companies organized in different ways depending on local or regional regulation. It is common to have a separation between the company managing the transmission network, which is called the *transmission system operator*, and the company managing the distribution, which is called *distribution system operator*. Both are referred to in this book as utilities. For the operation of the grid, utilities rely on one or more remote control systems managed from *utility control centers*.

From the telecommunication perspective, the term *telecommunication network* refers usually to the infrastructure (i.e., the set of nodes and links that provide the interconnection among different premises in the grid). Telecommunication networks, when complemented with other elements such as databases and operation and maintenance resources are referred to as *telecommunication systems*. These systems provide *telecommunication services*, which are used by the applications to implement the diverse smart grid functionalities. The services allow remote devices to connect within the needs of each application. *Protocols* define the rules for data transmission, reception, and processing to allow devices to understand each other properly; protocols define the conditions for establishing connections, the medium access control rules, the error correction procedures, the data representation, and the exchange sequence and formats. There are telecommunication-level protocols and application-level protocols.

If telecommunication services are not internally provided by utility private telecommunication networks, they are provided by public commercial telecommunication companies, to which we will refer as *telecommunication service providers*, also known as carriers or telecommunication operators.

As in the case of utility control centers, telecommunication networks are operated from *network management centers*.

The Book

This book is organized into nine chapters.

Chapter 1 provides a general view of the smart grids and how information and communication technologies (ICTs) are essential in their development.

Chapter 2 focuses on the description of the main concepts present in telecommunication technologies and networks.

Chapter 3 focuses on the description of the electric power system and its operation.

Chapter 4 describes the different smart grid applications, their nature, and their requirements.

Chapter 5 focuses on the most usual telecommunication technologies in the context of smart grids.

Chapter 6 discusses a telecommunication architecture for smart grid applications.

Chapter 7 is devoted to power line communication (PLC), as a telecommunication technology integrated in the grid. It includes several PLC use cases.

Chapter 8 is devoted to wireless (radio) technologies to effectively cover wide areas of the grid. It includes several wireless use cases.

Chapter 9 includes the guidelines to build the transition towards the smart grid. It includes the description of a use case integrating the different aspects described in the book.

A Final Word

The authors would like to express their gratitude to Iberdrola for its firm support and commitment to smart grids.

1

General View of the Smart Grids

Smart grid can be considered as the next step in the permanent evolution of the electrical grid since its inception. It is not a disruptive or innovative concept anymore, as it has been present in the industry for over a decade. However, it seems that right now there are an increasing number of utilities in their path to build real smart grids.

Considering this interest for the implementation of the smart grid and its large impact over grid operations, this chapter will clarify the scope of the smart grid concept, both from the utility standpoint and the customer perspective. The focus will be placed on the technical challenges, both for the electrical grid itself and the auxiliary technologies, specifically on telecommunications as one of the cornerstones of the smart grid realization.

1.1 What Is a Smart Grid?

1.1.1 Motivation

Evolution is a process of change in a certain direction. Any industry or business needs to evolve to adapt to the new context and take advantage of new opportunities.

Electrical grids have evolved since the first systems developed at the end of the nineteenth century and, despite a common misconception [1], have progressed to adapt to the growing needs of electrification, not only in terms of reach but also in terms of power consumption increase. The interconnected power grid has been mentioned as the "largest and most complex machine" on Earth [2], and it is clearly one of the greatest engineering achievements of twentieth century. However, it is also true that the technology cycles in utilities are longer than in other industries, that the substantial investments required by

1

some infrastructures prevent quick adoption of changes, and that risks associated with new technology must be well assessed.

The advances of electronics, computation, and telecommunications are continuously being applied to different aspects of our daily lives, and the utility industry is no exception. Many habits have changed due to the advent of information and communication technology (ICT), and many sectors have experienced important changes due to the adoption of ICT. There is a recent trend to apply the term "smart" to anything with which ICT is integrated (e.g., "smart cities"). The grid is no different and, since the beginning of the twenty-first century, any ICT, especially telecommunications applied to the electricity business, is marked as "smart grid." In fact, telecommunications can be considered the biggest challenge of smart grids due to the strict requirements imposed by the needs of the electric power system (e.g., latency) and the wide spread of the assets that are part of it (i.e., wherever there is a power supply).

The grid, or the electric power system, is not just one element. It is a whole complex system and it has many different components. It is also true that various parts of the grid are intrinsically disparate and present distinctive challenges. Thus, if we refer to smart grids, we need to understand where the grid "smartness" will be applied.

1.1.2 Formal Definitions

From the initial mentions to the smart grid concept in the 1990s, even if the smart grid was not a commonly agreed-upon expression (see [3–5]), the vision and first implementations of smart grids have evolved through different regions, countries, and utilities.

It is surprising to see how much these implementations differ among them, basically due to the different parts of the grid on which they are applied, the different applications and services from which they evolve, and the technical and economic constraints. This is so, despite the efforts of different institutions to make the vision and implementations converge through the compilation of standards, in an effort to make the smart grid environment one in which utilities can share practices, procedures, and grid and telecommunications network approaches. However, no two smart grid implementations are the same, probably because no two conventional grids are the same.

The smart grid begins with an idealistic vision of what is needed to make the electrical grid better. Anyone could probably draw the lines of a superior grid, through aspects such as:

- Modernization of grid infrastructure;
- Increase of the number of sensors and controls in the electricity system, managed from central systems;

- Monitoring and control of critical and noncritical components of the power system;
- Combination of bulk power generation and storage combined with distributed generation resources;
- Minimal environmental impact of electricity production and delivery;
- Automation of operational activities;
- Efficiency in the power delivery system and in the customers' consumption;
- Resilience of power supply, through the prevention of failures;
- Assurance of the power quality.

However, this set of objectives does not show how to accomplish them. Even more, as objectives, they can be achieved only to a certain extent within a set of constraints.

A smart grid is based on the application of ICT to the grid. The literature on smart grid offers different visions of the concept (see [5–13], from 2003 onwards). None of the definitions indicate how telecommunications are to be used to implement the smart grid. The common aspects of all definitions are summarized as:

- Presence of ICT in all parts and activities of the power grid, from the generation sources to consumer loads;
- Integration of ICT into the grid infrastructure and applications;
- Recognition of the complexity of the ICT concept, from hardware (electronics) to software and telecommunications;
- Alignment of ICT with the vision of a grid that needs it to achieve its objectives.

ICT, mainly understood as computing and electronic elements coupled with telecommunication networks, has historically been appreciated and used by the power utility industry. Grid elements have experienced an evolution from the times when they were monitored and operated at a local level. Centralized control centers progressively have taken control of most segments and elements of the grid. Automated systems, which take complex decisions based on data coming from different parts of the grid, have become a fundamental part of utility operations. Both the infrastructure (grid) and algorithms (intelligence) are fundamental for the smart grid, and the glue that integrates them is ICT.

Telecommunications are nothing new for utilities. Many telecommunication service providers (TSPs) today support their networks with the basic infrastructure (e.g., rights of way, ducts, poles, optical fiber cables) that is owned by utilities. A few of them are even spin-offs from utility companies. The history of utilities cannot be understood without the telecommunication networks they deployed for their own needs, due to the unavailability of public network services or their inability to comply with the requirements and mission-critical nature of electric assets.

Therefore, telecommunication networks are already an important part of utility companies' operations. However, the challenge of extending those networks to potentially millions of end-points, which are geographically dispersed over large service areas, is inherent to smart grid. The challenge is to grow the networks by possibly several orders of magnitude and in very diverse circumstances. Telecommunications for smart grid services cannot be systematically and cost-effectively provided by one single technology, but through a case-by-case seamless network mix of private and public telecommunication solutions. The optimal blend is different for each utility due to historical, economical, technical, market, or strategic reasons.

The ubiquitous presence of the smart grid cannot hide the fact that its implementation, or even the vision itself, is not yet mature. The problem is not the idea of a better grid, but the lack of widely accepted, publicly available references that can be taken as examples of good practices in smart grid implementation. This is partially derived from the historical evolution of companies and markets in various parts of the world, with different sizes, regulations, and targets, which mean that the interpretation of a "better grid" varies among utilities. Even the efforts to unify visions through standards and best practices have probably hidden the fact that we should possibly better discuss not of the smart grid as an ultimate goal, but of the "smarter grid" [5]. This evolution is in fact what the United States Energy Independence and Security Act summarized in 2007, with the "Statement of policy on Modernization of Electricity Grid" [14].

This book focuses on how the grid can be made "smarter" through the proper use of telecommunication technologies, networks, and services.

1.1.3   ICT and Telecommunications

ICT represents a broad concept that covers all technologies (hardware and software) that manage and process information data and transmit them through telecommunication networks.

ICT is certainly a term used extensively. The United Nations' International Telecommunication Union (ITU), the most prominent and widely recognized entity in the area, is defined as a specialized agency for information and communications technologies. National regulators, authorities, and companies have specific departments, strategies, and initiatives all related to the importance of ICT in our society.

The two main elements of ICT are information and communications. Information is a broad term which according to the International Electrotechnical Commission (IEC) is defined as "knowledge concerning objects, such as facts, events, things, processes, or ideas, including concepts, that within a certain context has a particular meaning." Further detail is given if the definition of information technology equipment is considered: "equipment designed for the purpose of (a) receiving data from an external source [...]; (b) performing some processing functions on the received data (such as computation, data transformation or recording, filing, sorting, storage, transfer of data); (c) providing a data output [...]".

Communications refers to the transit of information within networks, to move it from one point to another. The IEC defines communications as "information transfer according to agreed conventions." The term "telecommunications" is defined as "any transmission, emission or reception of signs, signals, writing images and sounds or intelligence of any nature by wire, radio, optical or other electromagnetic systems" in [15]. Thus, the concept of telecommunications applies more specifically to the contents in this book, and will be used from now on instead of the more generic *communications*. The authors think that the term *telecommunications* better addresses the subtleties and complexities in the context of smart grids.

1.2 Challenges of the Smart Grid

1.2.1 The Evolution of the Components of the Grid

The power grid is not a homogeneous and coherent single infrastructure but a large set of interconnected elements that altogether provide a service to its end customers. Although a full description of the electric power system will be given in Chapter 3, we will present its building blocks here to briefly discuss the level of integration of ICT today, along with the challenges for the future, in the different grid segments.

Figure 1.1 shows the major building blocks of a traditional electric power system, where the purpose is to deliver produced energy to the consumption end-points through transmission and distribution systems, the grid. Generation and consumption are matched in real time.

The traditional components of the electric power system are:

- Generation: This part produces the energy that is transported to the locations where it is needed. Generation is traditionally associated with

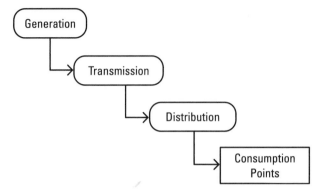

Figure 1.1   Building blocks of traditional electric power systems.

the big thermal, nuclear, and hydro plants, where energy transformation happens.

- Transmission: This segment is in charge of stepping generated voltage levels up, so as to minimize energy losses, and transport it long distances.

- Distribution: This segment takes care of driving conducted electric energy to the exact locations in which consumers are placed. Distribution systems are made of numerous, disperse assets with voltage levels that ultimately step down to the one being delivered to customers.

- Consumption points: These are the locations where energy is consumed. Electrification today is total in developed countries, while still more than 1 billion people live without electricity elsewhere [16].

Telecommunications are expected to penetrate further into the grid, as Figure 1.2 reflects.

At the same time, not any telecommunication network will be valid for implementing the smart grid, since different applications have divergent requirements that must be complied with. To achieve the vision of the smart grid [1], telecommunications need ideally to:

- Be bidirectional;
- Be wideband;
- Have low latency;
- Support massive, real-time data collection from possibly millions of end-points;
- Span vast, noncontiguous, and heterogeneous service areas;
- Be highly reliable.

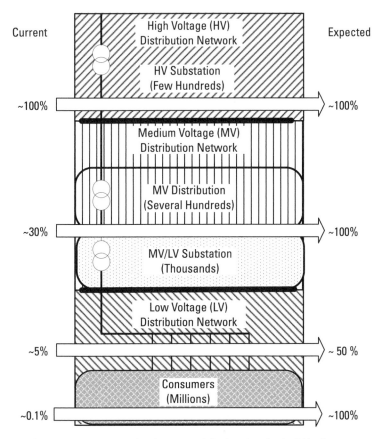

Figure 1.2 Expected telecommunications growth in the electrical grid [17].

• Be highly secure.

These requirements must be adapted and quantified for each application and segment of the electric power system that needs to support this ICT evolution.

However, the smart grid needs also to be understood as the evolution of the traditional power grid to accommodate the new building blocks that configure a modern grid (Figure 1.3). ICT makes this transformation possible.

The new elements that appear in the smart grid come with fresh challenges, some of which are mentioned next. ICT has a crucial role to make the elements integrate smoothly and realize their full potential.

1.2.1.1 Generation

The growth of electricity demand worldwide needs the untapping of new energy sources. Sources (e.g., mechanical, chemical, nuclear) need to be converted

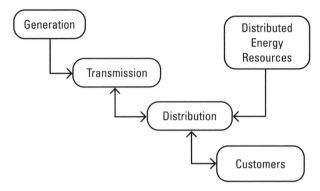

Figure 1.3   Building blocks of modern electric power systems.

into electric energy. This transformation involves a certain waste of energy that needs to be controlled and improved to achieve greater efficiency.

Environmental control (e.g., pollution management) is also a concern. Some old generation capacity will be retired and electric vehicles will enjoy wide adoption. However, while electric vehicle proliferation, reducing the pollution of traditional gasoline and diesel cars, will have a positive effect, this will need to be consolidated with clean energy sources that generate electricity.

Distributed generation (DG) will gain increasing importance in the operation and planning of power systems. Attached to many of these DG facilities, storage systems used for different applications are becoming popular in order to accumulate generated power. Storage systems include electric batteries, pumping water to reservoirs, compressed air, flywheel storage systems, fuel cells, super-capacitors, hydrogen, and others. Storage systems have implications in terms of maintenance, dynamic response, and charge/discharge cycle efficiency.

1.2.1.2   Transmission

Traditionally, transmission systems have been associated with keeping transmission line losses to a minimum. This is more important in the increasing electric demand context.

Additionally, growing congestion on the power flowing through the transmission lines must be controlled. When several generators connect to a long line, stability can be compromised if the power angle (voltage angle), that is, the voltage phase difference between the two ends of the line, increases.

Transmission systems need to control the damage from unexpected current flows induced on them, due to their greater length, and with origin in the space magnetic fields. These effects have been reported as the cause behind some outages.

Transmission systems need to control the power quality related to the frequency components additional to the fundamental frequency (50 or 60 Hz), known as harmonics, and caused by a variety of power electronics and switching components increasingly present in the grid.

1.2.1.3 Distribution

Automation, and specifically substation automation (SA) and distribution automation (DA), is concerned with the introduction of control systems in the operations of the distribution segment of the grid, as it is the most widespread part of the electric power system. The final objective is to bring higher quality of service and efficiency to the grid.

Automation is a key process that has been incrementally implemented in the grid. It started inside the substation and is being progressively extended towards the power lines and customers. In this expansion, the term advanced distribution automation (ADA) [18] has been used to emphasize the role of ICT. Self-healing, for example, is one part of the automation initiative; it will help distribution system operators (DSOs) to reduce the number of field interventions.

The appearance of DG in the supply chain is changing the electrical grid paradigm, as generation is no longer just at one end of the electric power system, but now also inserted into the distribution grid. This may create problems for protection schemes, as these systems are usually considered to work in specific types of grids, with certain topologies and impedances. When DG comes into play, the traditional assumptions (e.g., energy flow direction) do not necessarily apply and what it is supposed to increase reliability in the power system (i.e., DG) actually reduces it if the system is not properly designed. Safety considerations need to be taken into account too, due to islanding effects. In specific areas of the grid served by distributed generators, when power needs to be restored after a fault by field intervention, the grid might not be without voltage even if the fault remains, since voltage may be generated by the distributed source.

Some distributed sources may be needed in some power lines to maintain voltage levels. If this is the case, when a distributed generator is obliged to disconnect for the protection to recover the system, the utility will not allow the generator to reconnect until normal voltage levels are back, but this may never occur since the generator is needed for that purpose. This means that the traditional grid may only admit a certain number of distributed generators unless the smart grid implementation allows.

Another challenge is represented by DG inverters, which transform direct current (dc) power to alternating current (ac). Inverters depend on the magnitude and phase of the power line voltage in order to adjust themselves. When multiple inverters connect to the grid, the situation may be problematic if the voltage is just sensed from the grid.

1.2.1.4 Customers

Electrical grids have traditionally been oversized due to the difficulties to measure, understand, and modify customer energy use patterns. However, with the integration of ICT the situation is changing.

Customers have been traditionally using electricity as a commodity. They might need stimulus to raise their awareness and interest over their consumption patterns to get them to change. "Perceived usefulness" and "ease of use" concepts are key to make customers engage with any smart grid application.

Perceived usefulness goes further than the simple understanding of the consumer as an individual and focuses also on giving visibility to the customer over the whole process of energy management and environmental impact. "Ease of use" can be associated to intelligent appliances performing actions and aggregation of functions and taking some decisions on behalf of the customer under programmed criteria.

However, engaging consumers through new devices to trace their own consumption [as opposed to using their own personal computer (PC) or smart phone] or reducing their comfort with new consumption patterns could be difficult. Direct price orders to appliances and third parties optimizing energy use might have an interest for consumers as an alternative.

It has also been observed [19] that customers care about the control of energy for their own use, but care less about the management of any extra energy that they could generate and deliver to the grid, if they are also producers. This means that customer opinion needs to be considered before assuming any behavioral pattern.

Finally, customer visibility needs to be improved bidirectionally: consumers may want to see the impact of their actions; and utilities need to understand the consumption of their loads and when and how they are disconnected to improve internal operations.

1.2.2 Challenges of Telecommunication Design for Smart Grids

Telecommunication services are not independent from the networks supporting them and cannot provide anything that state-of-the-art networks cannot offer. This statement, although obvious, must be considered when designing smart telecommunication networks for smart grids.

Smart grid functions, from an ICT perspective, will be based on applications that are supported by telecommunications (e.g., to reach sensors and controls in distant power system areas). If the focus is on designing a system comprising applications and telecommunications with a top-down perspective (i.e., starting just by considering applications), there is the risk of getting requirements on the telecommunication networks and services that are either beyond the state of the art, out of the scope in terms of reasonable total cost of ownership (TCO), or simply not supported in already available telecommunication networks. Similarly, the risk on the other end is to consider the down-top approach as the method to design telecommunications for smart grid (i.e., to focus only on the limitations of today's networks, which might end up with a too-pragmatic smart grid design). The down-top approach would not stimulate the creation and evolution of telecommunication networks beyond their current possibilities. This is especially important for the smart grid, as utilities' investment cycles are not aligned with the business cases associated either to commercial telecommunication networks or to the fast evolution of ICT.

Thus, to efficiently design telecommunication networks and services for future-proof smart grids, it is important to consider that:

- Smart grids are not the ultimate and final evolution of electrical grids. On the contrary, smart grids are a path rather than a fixed objective.

- No single top-down or down-top approach can be used, if it is not in an iterative way, combining both until the solution fulfills the requirements.

- Requirements must be realistic and leading to achievable expectations in the evolution towards smart grid. Requirements based on simple, inexpensive, or traditional networks will produce design-limited solutions, unable to support the smart grid vision for long.

- The deployment of smart grids must be planned in phases. These phases are to be dictated by the needs of the electric elements that will be deployed in the power system and by the progressive evolution of the telecommunication solutions.

- No single solution exists for the "smarter grid": the difference of requirements in different parts of the network, the ubiquity of the solutions to deploy, the limitations on investments and expenditures, and the location of the premises to cover are so different that the mix of public/private and wireline/wireless networks is a must.

1.3 Grid Control as the Key for the Management of the Smart Grid

Control systems are probably the most important enabling element of a sustainable smart grid, as it is already the case in traditional electrical grids. Control is the way to guarantee that everything works, and that it will continue working. As [20] defined it, "the power grid is not only a network interconnecting generators and loads through a transmission and distribution system, but is overlaid with a communication and control system that enables economic and secure operation."

Control needs to happen in any relevant element of the electric power system. Control needs to be present in the generation plants to adjust the behavior of the generator. Control needs to be present feeding back the frequency offsets of interconnected grids (transport) to adjust generators. Control needs to be in the substations of the generation plants feeding the transmission segment grid, to be able to disconnect them when needed. Control is broadly used within the transmission substations, and inside them, to manage the different connections. Control needs to be extensively used in distribution grids to manage the connection with consumers and small-scale electricity producers and to provide resilience to the grid. Control needs to be in the hands of the consumers, to be able to actively engage in energy management.

Control systems include computation, electronics, and telecommunications; they form a multidisciplinary area. Specifically, telecommunications are relevant both in its local and remote perspective. Local telecommunications make the information exchange possible within premises. Remote telecommunications allow distant elements to communicate with utility control centers (UCCs).

Control has always been a part of the electric power system, and its evolution has been slow and incremental, following the slow change of operational procedures in power systems. Control was mainly based on human operators acting over a hierarchically designed system with data polling at low speed scanning rates (seconds) and limited feedback. Telecommunication systems have at some point in time been considered as the weak link [1] of DA.

The main role of telecommunications in control systems needing to cover geographically spread areas is to expand the scope, accuracy, and frequency of the information that needs to be gathered and to increase the capacity and networking capabilities of elements of the grid among them.

1.4   Consumer Engagement as the Key for the Transformation of the Smart Grid Model

The management of energy, the "good" supplied by the electric power system, has traditionally been restricted to the basic functions of transport and delivery. Electricity generation has traditionally been centralized, with little to no interaction with the customers.

However, the electric energy itself is fundamental, and the way that this electricity is consumed varies within the day and along the year. In the absence of storing capacity, generation systems must be able to support maximum peak demands. Even more, if consumption varies and the energy cannot be easily stored, generation plants must quickly adapt their production; this represents a challenge for some of the conventional generation technologies.

Another important aspect of energy is pricing. In many areas of the world, deregulation of power markets have brought along that energy prices are fixed by market dynamics, within certain rules often based on the existing generation plants and their constraints (e.g., capacity, investment, aging, nature).

In a traditional electric power system, the participation of small generation and consumers is not relevant. However, the situation is changing, as technology for small-scale generation is being made available. Small-scale storage systems are motivating a change in the electric system [21], and telecommunications are getting into the picture to allow for end-users and utilities to control consumption points at a finer scale in real time.

Some of the disruptive elements that impact consumer engagement, and will provoke a change in the way the system works today, are presented next:

- DG and distributed storage (DS): Wind power, solar power, geothermal power, biomass, and fuel cells are some of the resources available at a scale that allows the generation to be available for small producers and communities. This is a basic benefit of DG, as the generation source can be close to the consumption. From a situation in which electricity just flowed from large, identified generation facilities to one where production may exist anywhere and without previous knowledge of the system operators, the change affects the operational procedures of the grid.

- Demand side management (DSM): If production needs to match demanded energy, and the generation (and the transport grid) has to be dimensioned to cover demand peaks, it is necessary to control the demand of customers so as to achieve an optimal efficiency of the complete system. Customers might affect their consumption patterns based on the dynamic energy prices at specific times of the day. Different programs based on time of use (TOU) pricing, critical peak pricing (CPP), and real-time pricing (RTP) offer different rates to incentivize the use of energy away from certain parts of the day or days of the year. Some other specific programs exist under the name of demand response (DR). DR participation involves that either the utility or any other party (e.g., aggregator) may take control of different appliances to shut them off if and when needed, under a certain contract conditions.

Consumers are increasingly aware of their potential participation in the management of the energy that they consume [22], both for economic and sustainability reasons. The energy management of buildings, cities, and industries makes it possible for consumers to have a much greater control of what they consume from the utility, of how they can produce energy for self-consumption (and sell the excess to the system, possibly with price incentives), and of what each home appliance consumes. An expansion of this model will call for the deployment of microgrids, such that these systems may evolve to create small-scale electric systems, which could afford to be independent from large, public electric power systems, or as a controlled and modularized approach to the insertion of DG in the grid.

1.5　The Role of Telecommunications in the Smart Grid

Telecommunications are a broad field of knowledge that within the smart grid are keeping most of their complexity and diversity, while at the same time are constrained by the reality of the electrical grid.

1.5.1　Telecommunication Standards

A standard is basically a document that specifies, for a certain topic or related set of topics, agreed properties of manufactured goods, principles for procedure, and so forth. It is formally defined in ISO/IEC Guide 2:2004 as "a document, established by consensus and approved by a recognized body, that provides, for common and repeated use, rules, guidelines or characteristics for activities or their results, aimed at the achievement of the optimum degree of order in a given context."

Standards are fundamental for the evolution of any industry. Standards in ICT specifically address interconnection and interoperability, particularly important for open solutions that mix equipment and services from different suppliers in a competing market.

Standards have various origins (interest groups define requirements and engage in its development), are developed at the standardization bodies (all world regions, knowledge domains involved), are created in different ways (sometimes the standard inherits many of the characteristics of an industry solution; sometimes they are completely created from scratch), and finally have a different impact (market acceptance, longevity, and so forth). Standards can also have a different regional scope. However, most standardization bodies ultimately aim for global impact.

De facto standards are also important in the industry. Sometimes, due to the slowness of some standardization procedures, industry teams up to create solutions that become de facto standards, and sometimes they are eventually promoted to standards.

A quick reference is included to some of the most active and relevant standards organizations worldwide in the field of smart grid.

1.5.1.1 International Electrotechnical Commission (IEC)

IEC is, together with International Organization for Standardization (ISO) and ITU, an organization that, promoting technical collaboration among countries, prepares international standards for the world. Its knowledge domain is within electrotechnologies (i.e., electrical, electronics, and related technologies).

The IEC was founded in 1906 as a not-for-profit and nongovernmental organization. The IEC members are the National Committees, appointing delegates and experts. All IEC standards are fully consensus-based, with every country member having a vote.

The IEC has identified more than 100 standards within the scope of the Technical Committee (TC) 57 that are relevant for smart grids, placing the emphasis on some of them that are classified as core (see Table 1.1).

Other interesting standards are included in Table 1.2. They include all the standards marked as highly relevant (wind turbines, hydro, and electric vehicle have been removed):

1.5.1.2 International Telecommunication Union (ITU)

ITU is the United Nations agency specializing in ICT. ITU has both public and private sector members, so in addition to the 193 member states, ITU membership includes ICT regulators, academic institutions, and private companies. ITU was founded in Paris in 1865 and was organized in 1992 into three sectors, namely Radiocommunications (ITU-R), Standardization (ITU-T), and Development (ITU-D). The headquarters are based in Geneva, Switzerland.

A major role of ITU is to organize the World Radiocommunication Conference (WRC), held every three to four years, with the objective of reviewing and revising the Radio Regulations (RR) (international treaty governing the use of the radio-frequency spectrum and the geostationary-satellite and

Table 1.1
Core IEC Standard Series for Smart Grids

Topic	Reference	Title
Service-oriented architecture (SOA)	IEC/TR 62357	Power system control and associated communications; reference architecture for object models, services, and protocols
Common information model	IEC 61970	Energy management system application program interface (EMS-API)
Substation automation	IEC 61850	Communication networks and systems in substations
Distribution management	IEC 61968	Application integration at electric utilities; system interfaces for distribution management
Security	IEC 62351	Power systems management and associated information exchange; data and communications security

Table 1.2
Highly Relevant IEC Standard Series for Smart Grids

Topic	Reference	Title
Telecontrol	IEC 60870-5	Telecontrol equipment and systems. Part 5: Transmission protocols
TASE2: Telecontrol Application Service Element	IEC 60870-6	Telecontrol equipment and systems. Part 6: Telecontrol protocols compatible with ISO standards and ITU-T recommendations
Distribution Line Message Specification	IEC 61334	Distribution automation using distribution line carrier systems
Metering	IEC/TR 62051	Electricity metering
Metering	IEC 62052	Electricity metering equipment (ac): general requirements, tests and test conditions
Metering	IEC 62053	Electricity metering equipment (ac): particular requirements
Metering	IEC 62054	Electricity metering (ac): tariff and load control
Metering	IEC 62058	Electricity metering equipment (ac): acceptance inspection
Metering	IEC 62059	Electricity metering equipment: dependability
Companion Specification for Energy Metering	IEC 62056	Electricity metering: data exchange for meter reading, tariff, and load control

nongeostationary-satellite orbits). The WRC is key to the coordination and allocation of frequencies at a worldwide level, and the process is a heavy one, as the implications of changes impact the established situation and its future evolution in a globalized economy.

Within the smart grid, ITU-T created a focus group to track smart grid activities and identify impacts and opportunities. The relevant smart grid-related ITU recommendations will be mentioned throughout the book.

1.5.1.3 Institute of Electrical and Electronics Engineers (IEEE)

The IEEE is an association dedicated to advancing innovation and technological excellence. The IEEE claims to be the world's largest technical professional society serving professionals involved in all aspects of the electrical, electronic, and computing fields and related areas of science and technology.

The IEEE's origins date back to 1884 in the United States. Since its foundation, IEEE has developed societies (38 to this day) from the professional groups of the former constituents. The two societies more related to the smart grid activities are the IEEE Communications Society and the IEEE Power & Energy Society. The IEEE references more than 100 standards related to the smart grid, and many of them will be mentioned along this book.

1.5.1.4   European Telecommunications Standards Institute (ETSI)

The ETSI is one of the three bodies [together with Comité Européen de Normalisation (European Committee for Standardization, CEN) and Comité Européen de Normalisation Electrotechnique (European Committee for Electrotechnical Standardization, CENELEC)] officially recognized by the European Union (EU) as European Standard Organizations (ESOs). ETSI is focused on European standards for ICT, but pursuing a global applicability. ETSI is a not-for-profit organization with more than 800 member organizations worldwide in 64 countries, working under the principles of consensus and openness in the standardisation process. The ETSI works closely with the National Standards Organizations (NSOs) in the European countries, and in particular, all the European Standards (ENs) become national standards of the different European member states.

The ETSI was created in 1988. The ETSI works very close to the EU institutions. In the smart grid scenario, the European Commission has issued several mandates to CEN, CENELEC, and ETSI to develop standards for smart grids (M/490), smart metering (M/441), and charging of electric vehicles (M/468). In particular, to drive the coordinated answer of ESOs to M/490, a CEN-CENELEC-ETSI Smart Grid Coordination Group was created. This group created a set of reports approved by the three bodies at the end of 2014 that cover a reference architecture model and a list of all necessary standards for the smart grid. The standards are a compilation of the work of many other institutions globally [23]. A similar organization was built for the M/441 Smart Metering mandate, with the Smart Meters Coordination Group.

1.5.1.5   American National Standards Institute (ANSI)

The ANSI is a private nonprofit organization with the mission to facilitate the standardization and conformity assessment in the United States, for the better performance of the internal market, and the strengthening of U.S. market position at a worldwide level. The ANSI is also the official U.S. representative to the ISO and indirectly to the IEC. The ANSI develops third-party accreditation services to assess the competence of organizations certifying products and personnel.

The ANSI was founded in 1918 and does not develop American National Standards (ANSs) by itself. Rather on the contrary, it provides interested U.S. parties a framework to work towards common agreements, within the principles of due process, consensus, and openness. Hundreds of ANSI-Accredited Standard Developers in the private and public sectors develop and maintain American National Standards. The ANSI label may be used when the organization producing the standard meets ANSI requirements; this is the case with the IEEE.

1.5.1.6   Other Standardization Bodies and Interest Groups

The smart grid initiative is a global one. It would be too cumbersome to name all the relevant regions of the world (e.g., Japan, Korea, China, India) and the different agencies, regulators, or institutes. For the sake of simplicity, we will refer to other European and American bodies.

The National Institute of Standards and Technology (NIST) is a nonregulatory federal agency and one of the nation's oldest physical science laboratories now within the U.S. Department of Commerce. The NIST was founded in 1901 to create a first-rate measurement infrastructure to support U.S. industrial competitiveness. In the smart grid domain, the NIST established the Smart Grid Interoperability Panel (SGIP) to coordinate standards development in this area. The SGIP is composed of private and public sector stakeholders, and in 2013 evolved as a nonprofit private-public partnership organization, SGIP 2.0, with NIST continuing with an active role. The SGIP intended to accelerate standards harmonization and advance into the implementation and interoperability of smart grid devices and systems.

The Conseil International des Grands Reseaux Electriques (International Council on Large Electric Systems, CIGRE) is an international nonprofit association whose objective is to promote collaboration with experts from all around the world by sharing knowledge and joining forces to improve electric power systems today and in the future. CIGRE's central office is in Paris, France, and was founded in 1921. CIGRE works with experts in Study Committees overseen by the Technical Committee. The Study Committee SC D2 Information Systems and Telecommunication includes within its scope the study of the new ICT architectures to control bulk power systems (e.g., smart meter, smart grid, intelligent grid).

The Electric Power Research Institute (EPRI) is an independent, nonprofit organization, bringing together scientists and engineers, and experts from academia and industry, to address the challenges in all the aspects of the electricity domain, through the research, development, and demonstration of different initiatives within the interest of the sector. The EPRI was stablished in 1972, as a consequence of the growing U.S. concern on the dependence on electricity, which crystalized in a U.S. Congress initiative. Although its origin was in the United States, it extends to more than 30 countries with members from more than 1,000 public and private organizations. In the smart grid domain, the Intelligrid concept as the architecture for the smart grid of the future is one of the best known contributions.

The Utilities Telecom Council (UTC) is a global trade association with the purpose of creating a favorable business, regulatory, and technological environment for companies that own, manage, or provide critical telecommunication systems in support of their core business (e.g., utilities including electricity utilities). It was founded in the United States in 1948 with the initial focus of getting radio spectrum allocations for power utilities, and it is now focused on ICT solutions for its members all around the world.

1.5.2 Specifics of Electrical Grid Infrastructure

One of the most common errors when approaching the operational telecommunication needs of utilities is to forget the nature of the grid infrastructure. It is indeed fundamental to understand not only the service requirements in terms of the common telecommunication parameters, but to identify the characteristics of the locations where these services need to be provided.

Being the grid infrastructure present on most inhabited areas, the first aspect to notice is that if all electric assets are to be serviced (e.g., smart meter access), the reach of the network has to be larger than the one that typical TSPs provide commercially. Even if telecommunication service is increasingly being considered as a basic one and universal service obligations [24] are imposed on incumbent telecommunication operators worldwide, there is still a gap between the locations with electricity service available, and those with access to telecommunications. Even if the gap seems small in percentage terms, it is significant when referring to territory.

If we focus on telecommunications and follow the traditional classification of urban, suburban, and rural areas, typically we can expect a better coverage (in terms of variety of sources, technologies, bandwidth, throughput, and latency) in urban and suburban scenarios. In the case of electric premises, however, they tend to be hidden or placed where they cannot be easily seen or do not interfere with public activities. Substations tend to be outside the inhabited areas and/or below ground. Meters are located on ground floors in communal areas, in cellars or inside boxes that in many cases make radio frequency (RF) propagation difficult. Rural areas are even more difficult, as they are not typically attractive for commercial TSPs, and are usually underserved by them.

The other aspect that is usually not clearly understood is the need to prepare telecommunication products (devices, usually) to perform in aggressive environmental conditions. Electric premises do not belong to typical industrial scenarios, but they include strict electromagnetic compatibility (EMC) requirements due to the special nature of electricity in locations where voltage is measured in kilovolts. This is especially true where substations are served by metallic wires intended for telecommunications with another premise; in these cases, ground rise potential effects must be taken very seriously into consideration, as people safety may be affected.

Thus, very often commercially available solutions (services and equipment) cannot be adopted for many of the telecommunication needs of the grid, and as a consequence the evolution of a grid towards a smart grid cannot happen without the adaptation of existing products and the deployment of new or enhanced networks.

1.5.3 Legacy Network and Grid Element Integration

Grid operation applications and technologies using telecommunication technologies already exist in utilities. Supervisory control and data acquisition (SCADA) systems, used for distribution grid control, can be presented as an example of the situation. These systems were built in the last decades largely using plain analog telephone circuits, exploited with the help of modems. On the physical media side, both voice wireline and wireless networks have supported the service delivery to reach the remote terminal units (RTUs). Regarding the interfaces, baseband to serial modems made the transit of digital signal over the analog channels possible. Thus, RTUs either implemented internal modems or relied on external ones and configured products that were extensively delivered to the field and supported by dedicated telecommunication networks that still exist to this day.

The problem with legacy contexts is financial and technical. The kind of electricity infrastructure attached to these systems is affected by long amortization periods that do not admit short-term changes. On the technical side, the problem is twofold. First, vendors providing products for the RTUs usually have their origins (specially the smaller ones) within the electric industry, and their primary skills are within that industry. Thus, the speed at which they can adapt to new telecommunication technologies is limited, and this may hamper product evolution. Second, different telecommunication networks present specific difficulties:

- In the case of wireline, there are problems both in private and public telecommunication networks. If the telecommunication network is private and consists of copper pairs, the improvement of the service is limited by the evolution of the available network. Even if all digital subscriber line (xDSL) systems have allowed the reuse of these assets, xDSL obtains good results in short distances, which is usually not the case where utilities own these copper assets. If the network is public, the telecommunication operators have long ago decided not to invest further in their own analog networks.

- In the case of wireless, the unavailability of digital radio networks some decades ago implied that each utility and vendor ecosystem took their decisions in terms of how to better develop private radio networks to allow SCADA systems to take control of RTUs over their service areas. Both the radio telecommunication networks and the protocols on top of them were proprietary with a few exceptions. Many of these decisions led to the development of protocols that, on top of an analog medium, would get some sort of medium access control (MAC) to allow the SCA-DA master station to communicate with the RTUs. As a consequence, existing channels cannot be reused by different vendors.

Therefore, the problem of legacy technology and elements coexistence needs to be acknowledged in the electrical grid.

The fundamental difference with other industries is that, while in the public/commercial telecommunication sector, any evolution or change can be justified with a business plan to make a profit (a new service that can be commercialized based on the extended capabilities of the new technology), in the electric utility sector (e.g., distribution of energy kept as a regulated activity) there is neither a business case nor a clear and linear relationship between the change in telecommunication technology and a better quality of service. "Green-field" scenarios for utilities cannot be easily justified, and networks and elements with different capabilities and operational constraints will need to coexist for long periods.

1.5.4 Single-Network Vision

There are two aspects that need to be highlighted under the vision of a single telecommunication network supporting the smart grid services and applications.

The first aspect refers to the traditional trend of developing single-service oriented networks, that is, networks that due to their origin, supported services, performance, or design constraints are only able to carry one single service or a set of reduced specific services tailored to a single application.

The telecommunication services needed for the smart grid are only reasonable from the economic perspective when they are provided by telecommunication networks that do not need to be refurbished whenever there is a need to add a new service. The investment associated to network development is huge, as the assets usually spread all over the service area, and any duplication of telecommunication network implies important costs. Thus, there is a clear economic rationale behind multiservice networks. On the side of the difficulties, designing and developing multiservice networks need a clear understanding of the needed future services, as well as a higher up-front investment.

The second aspect is related to the need to control and manage the telecommunication services provided by different combinations of technologies at the open system interconnection (OSI) layers (see Chapter 2) and origin of the telecommunication services (public or private), in a uniform way.

The implication is that there will be potentially many different devices, networks, and services to be monitored and managed. These elements will be coming from different vendors and will have their own management platforms.

In the case of the services, different TSPs will offer different access levels to the service status for the utility to manage in real time. The poorest case of the utility vision of the service will be that of the pure administrative access for billing and contract service-level agreement (SLA); the documented agreement between the service provider and customer identifying service targets. In this context, the utility will be able to manage each one of the layers of the network or the particular service, but will not be able to have the desirable end-to-end management capabilities of the services the network is providing.

1.5.5   Broadband Expansion and Distributed Intelligence

The smart grid concept assumes a superabundance of sensors and controls all around the grid. The sensors will provide all the data needed to understand the status and performance of the different grid elements, and the controls provide the ability to act over those elements. With the easy access to low-cost processing capability some of this control can be executed locally, based on some algorithms and logic; the status and results will be reported to the utility control centers (UCCs).

However, this way of operating is a result of the evolution of the control systems when telecommunications were a scarce low-performing resource. A system full of small islands, isolated and unable to exchange data, forced to take decisions based on their own local view of the situation, is less attractive than a fully interconnected grid with broadband capabilities that allow the algorithms, the operational logic, and the human control reside where it is more appropriate for the utility, adapting to the specific constraints and roadmap of each one.

Thus, telecommunications are a driver not only to enable the smart grid, but also to help to organize operations, applications, and infrastructure development, adapting to the needs of each utility. Probably restrictions of the past, giving as a result concepts we keep as valid today, need no longer to be kept as axioms. Intelligence will be centralized or distributed not because of the limitations of the telecommunications, but based on the decision of the utility.

1.5.6   Network Design

Telecommunication network design is the initial step to guarantee smart grid applications performance. Performance, scalability, reliability, and security are to be considered in a network design.

1.5.6.1   Network Performance

The early days of telecommunications employed dedicated circuits for specific purposes. Progressively, these dedicated assets were shown not to be scalable and networks emerged. When resources are shared inside networks, and the users or services may have different behavioral patterns, performance needs to be planned and controlled.

Network performance can be defined as in ITU-T Recommendation E.800, as "the ability of a network or network portion to provide the functions related to communications between users." Performance is critical for applications to deliver their expected benefits. The different telecommunication services demanded by smart grid applications need to be compliant with certain performance objectives that are built around. As an example:

- Throughput: The maximum data rate where no packet is discarded by the network. Throughput is usually measured as an average quantity, with control over the peak limits.

- Latency: The time it takes for a data packet to get from one point to another. Latency is usually defined as a requirement below a certain limit.

Telecommunication networks need to provide specified quality for the services they need to deliver. This is the quality of service (QoS), a key element of any telecommunication network, both as a design parameter and as the user perception once the network is deployed.

ISO 8402 defined the QoS as "the totality of characteristics of an entity that bear on its ability to satisfy stated and implied needs." ISO 9000 defined it as the "degree to which a set of inherent characteristics fulfils requirements." ITU-T Recommendation E.800 defined QoS as "the collective effect of service performance which determine the degree of satisfaction of a user of the service." This recommendation brings together QoS and network performance, but clearly establishes that the QoS is not just dependent on network performance, but on non-network performance aspects as well (Figure 1.4). It also mentions end-to-end QoS, encompassing all service aspects.

QoS can be used as a tool to define telecommunication services for users. ITU-T Recommendation G.1000 referred to ETSI ETR 300 series,

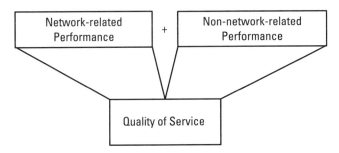

Figure 1.4   QoS is affected by network and non-network-related aspects performance.

summarizing the different service functions (sales, service management, technical quality, billing processes and network/service management by the customer) and quality criteria with each of them (speed, accuracy, availability, reliability, security, simplicity, and flexibility).

1.5.6.2 Network Scalability

The main reason to create networks at the early stages of telecommunication history was the need to cope with the growing number of users in the network. The classical example of this was the Plain Old Telephone Service (POTS), which evolved into a system with telephone exchanges to avoid an excess of assets (telephone lines) and progressively into a hierarchy to make the network escalate to the maximum number of users needed worldwide.

Network growth can be the result of new end-points being added to a network, or extended bandwidth being required. It should not be the result of harder delay constraints, as this could involve a design change.

Thus, any network design should be made considering the performance (throughput and latency) of the services it needs to deliver, considering that it needs to accommodate the evolution in the coming years, as well as of other services and end-points. The design should be implemented using and offering interfaces in which the industry invests heavily (Ethernet/IP) and will not require a redesign whenever new services, end-points, or new coverage areas are added. The design needs to consider the modularity of the network elements in terms of number of interfaces and switch fabric capabilities as well.

The smart grid concept is not yet fixed, and probably will differ in many regions and countries. As ICTs come closer to utilities, new applications will appear as a result of the new possibilities that could be identified. Thus, the target of the smart grid will still be a moving one for the coming years, and scalable designs are needed to guarantee that new ideas and applications are not hampered by limited designs that may need high investments to be adapted.

1.5.6.3 Network Reliability

Networks and systems are often referred to as reliable or dependable, and we mention their resilience and measure their availability.

Something is reliable if it is dependable (i.e., it is consistently good in quality) and gives the same result in successive attempts. Something is resilient if it is capable of recovering from or adjusting easily to misfortune or change; it needs to be able to return quickly to a previous good condition. Something is available, if it is ready for immediate use.

ITU-T Recommendation E.800 defined availability, reliability, and dependability, in relation to service quality characteristics:

- Availability as the ability "of an item to be in the state to perform a required function at a given instant of time or at any instant of time within a given time interval, assuming that the external resources, if required, are provided";

- Reliability as "the probability that an item can perform a required function under stated conditions for a given time interval";

- Dependability as the "performance criterion that describes the degree of certainty (or surety) with which the function is performed regardless of speed or accuracy, but within given observation interval."

The IT world has also definitions that take us closer to the real implications. Reference [25], specifically for IT systems, referred to reliability as "a measure of the percentage uptime, considering the downtime due only to faults." Availability is in the context of the same reference "a measure of the percentage uptime, considering both the downtime due to faults and other causes (e.g., planned maintenance)." Thus, one system can be more reliable but less available than the other, if planned maintenance (changes, upgrades, and so forth) takes time. Resilience is related to all the actions contributing to achieve the necessary reliability.

The European Union Agency for Network and Information Security (ENISA) has defined resilience as "the ability of a system to provide & maintain an acceptable level of service in the face of faults (unintentional, intentional, or naturally caused) affecting normal operation" [26]. Resilience, by contrast, is defined as "the ability of an entity, system, network or service to resist the effects of a disruptive event, or to recover from the effects of a disruptive event to a normal or near-normal state." Thus, while reliability focuses on what is in the system not to fail, resilience focuses on recovering when something fails.

Availability objectives are common practice in telecommunications. ITU-T Recommendation G.827 established them for general data circuits; ITU-R F.1703 recommendation fixed "the availability objectives for real digital fixed wireless links used in 27,500 km hypothetical reference paths and connections."

There is not yet a commonly agreed framework of reliability targets for the smart grid services (see Table 1.3). It is broadly assumed that the most strict services might need 99.999% availability, that translates into (100% − 99.999% = 0.001%) unavailability.

Network design must count on aspects such as carrier-grade or carrier-class network elements (redundancy in vital parts of the devices: power supply and switching parts), reliable design of network links (e.g., design of radio links), protection of paths, license spectrum usage (that is legally protected not be interfered), terminals with appropriate mean time between failure (MTBF), and battery backup in interruptible-prone locations.

Table 1.3
Availability Levels

System Type	Unavailability (minutes/year)	Availability (%)
Unmanaged	52,560	90
Managed	5,256	99
Well managed	525.6	99.9
Fault tolerant	52.56	99.99
Highly available	5.256	99.999
Very highly available	0.5256	99.9999
Ultra highly available	0.05256	99.99999

Source: [27].

1.5.6.4 Network Security

Since the early dates of the open-channel radio emissions (which did not protect from eavesdropping), security aspects have become increasingly important. As an example of this importance, North American Electric Reliability Corporation (NERC) has developed and enforced regulations that are a clear demonstration of the importance of security in electric utilities.

Parallel to network design, security cannot be simply accomplished with a collection of tools or recommendations. It is rather a process in which security is progressively incorporated to all the processes and assets of the utility, so as to comply with externally or internally imposed procedures. Security is not just a technological concern, but a procedure and process-driven approach.

Networks need to consider security in the data flows, but also in the networks themselves, and the access to network elements. Data flows need to be encrypted, but they also need to be controlled in its routes; even if telecommunication services share common networks, this does not mean that services are not independent and isolated.

Access control should be executed over the terminals and network elements connecting to the network to guarantee both that the hardware is controlled and that the connecting element is the one it declares. These elements also need to verify that the systems to which they connect are genuine.

Regarding the operation of the network, all individuals connecting to the network, network elements, or terminals should be just granted access to the element or part of the network that they need to operate and for the time and purpose they need. Operational procedures become more complex as a result.

1.5.7 Private Networks Versus Public Services

The emergence of electricity grids and telecommunications are parallel stories. Initially, they were two very similar disciplines, up to a point in which specialization made them follow different paths. Since the early days, utilities had the need to develop telecommunication networks for the operations of remote assets, where no telecommunications were present. Examples of this are the early powerline carrier developments that have evolved into the current power line carrier (PLC) technologies.

Progress in ICT has brought along the generalization of telecommunication services, mainly at residential level. It is not difficult to find today generic services that seem suitable for the electrical grid applications. Technical misconceptions, underestimation of requirements, overestimation of capacity capabilities and coverage, and the wrong assumption that telecommunications will not be integral part of the electricity services have made existing commercial telecommunications appear to some as the straightforward solution to the telecommunications for smart grid services and applications.

There are some reasons favoring the assumption that public/commercial telecommunication services are valid for smart grids:

- Public/commercial telecommunication services are fully operational. This is, in the places where the service is effectively delivered and in the absence of incidents, the applications of the utility, "work".

- The immediate investment in public/commercial telecommunication services is lower than the cost of deploying a network for the provision of services.

- It may be assumed that the existing competition in telecommunications domain might render economic benefits in the cost of the services.

- The assumption that telecommunication services are readily available in the existing public networks might make utilities quickly and timely deliver services.

- The lack of telecommunication expertise within some utilities (especially the smallest ones).

- In the case of radio networks, expensive or difficult access to spectrum is another reason that might make utilities rely on TSPs.

However, there is evidence that public networks are not well suited for mission-critical services, or those that require availability or QoS to be higher than the average. In [28, 29] some findings and lessons learned are summarized:

- Standard services offer a variable degree of redundancy.

- Redundancy of services from different telecommunication carriers in the same location cannot be assumed, as many back-up systems are provided over the same physical media (e.g., the same cable).

- The "last mile" (the segment from the local exchange to the customer premises) is usually the single point of failure even in networks with a high level of resilience.

Additionally, the lower initial investment of TSP's services cannot hide other longer-term costs. The TCO of the solution (capital expenditure and operational expenses) must be taken into account:

- It is important not only to consider the up-front costs, but the recurrent operational expenses, that are usually higher as the data exchange increases (e.g., if a more intensive use or new applications are added).
- Not only is the cost of the first installation of the service important, but also the eventual need to replace the telecommunication terminal (e.g., change of technology; cycles in mass-market technologies are quicker that the ones associated to utilities) or to move to another TSP for commercial reasons:
  - Smart grid assets in utilities are not always easy to locate and access, and many times require that the visit to the premise is performed by more than a single individual for safety reasons.
  - Not all the telecommunication terminals can be reused from one TSP to the other, despite the efforts in standardization (e.g., some satellite-based services).

Based on this evidence and local findings, there is a trend in utilities to prefer privately owned telecommunication networks to deliver the services needed for smart grids. The reasons are:

- The need to design and deploy networks to effectively cope with the electricity service commitments of the utility:
  - The utilities need to operate their applications for operational purposes, for example, SCADA systems or teleprotections are included in the codes of practice of the utilities. The performance of the applications is hampered if the telecommunication services performance (bandwidth, latency, and others) are not achieved and services are disrupted. This noncompliance may result in penalties that the utilities need to face, as well as physical assets losses.
  - The utilities need to comply with the safety and security conditions imposed by both the regulators and the nationwide authorities. Non-

compliance will imply penalties and even the revocation of the operation license.

- The need to integrate the management of telecommunication services with the electricity grid control. UCCs manage the electricity grid through different applications and systems that they own. The integration of third-party telecommunication services provides less control when the networks do not belong to the utility.

Some other factors derived from [30] are:

- Telecommunication networks for public/commercial services do not generally offer SLAs that match the needs of many of the mission-critical nature of many of the services needed for smart grids.
- Public commercial networks are subject to regulation from the national authorities, and even if they are to be changed, the national authorities need to authorize it and study its impact in their local markets.
- Public networks do not in general offer priorities in the network over the rest of the users. If these networks need to be adapted, both standards and deployed infrastructure should be modified.
- Public telecommunication networks do not reach everywhere. Scarcely populated areas are not well served with telecommunications for economic reasons, and in the case of radio services, penetration of the signals where the service is needed is not always possible with the existing network (e.g., assets below ground).

References

[1]   Goel, S., S. F. Bush, and D. Bakken, (ed.), *IEEE Vision for Smart Grid Communications: 2030 and Beyond*, New York: IEEE, 2013.

[2]   Schewe, P. F., *The Grid: A Journey Through the Heart of Our Electrified World*, Washington, D.C.: Joseph Henry Press, 2007.

[3]   Haase, P., "Intelligrid: A Smart Network of Power," *EPRI Journal,* No. 2, 2005, pp. 26–32. Accessed March 21, 2016. http://eprijournal.com/wp-content/uploads/2016/02/2005-Journal-No.-2.pdf.

[4]   Carvallo, A., and J. Cooper, *The Advanced Smart Grid: Edge Power Driving Sustainability, Second Edition*, Norwood, MA: Artech House, 2015.

[5]   Gellings, C.W., *The Smart Grid: Enabling Energy Efficiency and Demand Response*, Lilburn, GA: The Fairmont Press, 2009.

[6]   International Low-Carbon Energy Technology Platform, *How2Guide for Smart Grids in Distribution Networks*, Paris: International Energy Agency (IEA), 2015.

[7]   European Commission, *Standardization Mandate to European Standardisation Organisations (ESOs) to Support European Smart Grid Deployment*, Brussels, 2011.

[8]   Korea Smart Grid Institute. "Korea's Jeju Smart Grid Test-Bed Overview," 2012. Accessed March 27, 2016. http://www.smartgrid.or.kr/10eng3-1.php.

[9]   The Climate Group and Global eSustainability Initiative (GeSI), *SMART 2020: Enabling the Low Carbon Economy in the Information Age*, 2008. Accessed March 27, 2016. http://www.smart2020.org/_assets/files/02_smart2020Report.pdf.

[10]   Adam, R., and W. Wintersteller, *From Distribution to Contribution: Commercializing the Smart Grid*, Munich, Germany: Booz & Company, 2008. Accessed March 27, 2016. http://www.strategyand.pwc.com/media/uploads/FromDistributiontoContribution.pdf.

[11]   Miller, J., "The Smart Grid – An Emerging Option," *Fall 2008 Conference: Responding to Increasing Costs of Electricity and Natural Gas of the Institute for Regulatory Policy Studies (IRPS)*, Illinois State University, Springfield, IL, December 10, 2008. Accessed March 27, 2016. http://irps.illinoisstate.edu/downloads/conferences/2008/IRPSMillerFinalPPT120808.pdf.

[12]   European Commission – Directorate-General for Research – Sustainable Energy Systems, *European Smart Grids Technology Platform: Vision and Strategy for Europe's Electricity Networks of the Future*, Luxembourg: Office for Official Publications of the European Communities, 2006. Accessed March 27, 2016. ftp://ftp.cordis.europa.eu/pub/fp7/energy/docs/smartgrids_en.pdf.

[13]   U.S. Department of Energy, Office of Electric Transmission and Distribution, *GRID 2030: A National Vision for Electricity's Second 100 Years*, 2003. Accessed March 27, 2016. http://energy.gov/sites/prod/files/oeprod/DocumentsandMedia/Electric_Vision_Document.pdf.

[14]   U.S. Congress, "Title XIII – Smart Grid, Section 1301 Statement of Policy on Modernization of Electricity Grid," *Energy Independence and Security Act*, 2007.

[15]   International Telecommunication Union – Radiocommunication (ITU-R), *Radio Regulations Edition 2012*, Geneva, 2012.

[16]   International Energy Agency (IEA), "Modern Energy for All," 2015. Accessed March 21, 2016. http://www.worldenergyoutlook.org/resources/energydevelopment/#d.en.8630.

[17]   Mott MacDonald Group Ltd., "Utility Requirements for Smart Grids," *Workshop Utilities and Telecom Operators: Collaboration for the Future Energy Grid?*, Brussels, March 19, 2014.

[18]   Goodman, F., and M. Granaghan, "EPRI Research Plan for Advanced Distribution Automation." *IEEE Power Engineering Society General Meeting*, 2005.

[19]   Bush, S. F., *Smart Grid: Communication-Enabled Intelligence for the Electric Power Grid*, New York: Wiley-IEEE Press, 2014.

[20]   Tomsovic, K., et al., "Designing the Next Generation of Real-Time Control, Communication, and Computations for Large Power Systems," *Proceedings of the IEEE*, Vol. 93, No. 5, 2005, pp. 965–979.

[21]   Bronski, P., et al., *The Economics of Grid Defection*, Basalt, CO: Rocky Mountain Institute, 2014.

[22]   eBADGE Project, "Project Description," 2012. Accessed March 21, 2016. http://www. ebadge-fp7.eu/project-description/.

[23]   European Committee for Standardization (CEN) and European Committee for Electrotechnical Standardization (CENELEC), "Smart Grids," 2016. Accessed March 21, 2016. http://www.cencenelec.eu/standards/Sectors/SustainableEnergy/SmartGrids/Pages/default.aspx.

[24]   Directorate for Financial and Enterprise Affairs, *Universal Service Obligations*, Paris: OECD, 2003. Accessed March 21, 2016. http://www.oecd.org/regreform/sectors/45036202.pdf.

[25]   Scott, R., "The Differences Between Reliability, Availability, Resiliency," *ITBusiness.ca*, 2006. Accessed March 21, 2016. http://www.itbusiness.ca/news/the-differences-between-reliability-availability-resiliency/8556.

[26]   Gorniak, S., et al., "Enabling and managing end-to-end resilience," Heraklion, Greece: European Network and Information Security Agency (ENISA), 2011. Accessed March 21, 2016. https://www.enisa.europa.eu/activities/identity-and-trust/library/deliverables/e2eres/at_download/fullReport.

[27]   Gellings, C.W., M. Samotyj, and B. Howe, "The Future's Smart Delivery System [Electric Power Supply]," *IEEE Power and Energy Magazine*, Vol. 2, No. 5, 2004.

[28]   Centre for the Protection of National Infrastructure (CPNI), *Telecommunications Resilience Good Practice Guide Version 4*, 2006. Accessed March 21, 2016. https://www.cpni.gov.uk/documents/publications/undated_pubs/1001002-guide_to_telecomms_resilience_v4.pdf.

[29]   Lower Manhattan Telecommunications Users' Working Group, *Building a 21st Century Telecom Infrastructure*, New York, 2002. Accessed March 21, 2016. http://www.nexxcomwireless.com/wp/wp-content/uploads/2013/10/Building-a-21st-Century-Telecom-Infrastructure-Lower-Manhattan.pdf.

[30]   Forge, S., R. Horvitz, and C. Blackman, *Is Commercial Cellular Suitable for Mission Critical Broadband?*, European Commission DG Communication Networks, Content & Technology, 2014. Accessed March 21, 2016. http://www.era.europa.eu/Document-Register/Documents/FinalReportEN.pdf.

2

Telecommunication Concepts for Power Systems Engineers

Readers will usually have a stronger background in either telecommunications or power systems. This chapter will give a general view of useful concepts specific to telecommunication systems. Telecommunications have been defined by the International Electrotechnical Commission (IEC) and the International Telecommunication Union (ITU) as "any transmission, emission or reception of signs, signals, writing, images and sounds or intelligence of any nature by wire, radio, optical or other electromagnetic systems."

2.1 Concepts

2.1.1 Telecommunication Network

A telecommunication network is a set of devices designed to exchange information remotely, typically in real time and in most cases bidirectionally (radio broadcasting is an example of unidirectional telecommunication network). Networks, when accompanied by nontelecommunication elements (databases, information systems, and so forth), are considered systems. There are usually several technologies involved in a complete telecommunication system.

Devices in a bidirectional telecommunication system can act as transmitters and as receivers, alternatively or simultaneously. If it is alternatively, transmission is called half-duplex. If it is simultaneously, transmission is called full-duplex.

Devices are connected through transmission media such as copper cables, optical fibers, and free space. Transmission media together with the network nodes constitute the network. For example, intelligent electronic devices (IEDs) or remote terminal units (RTUs) installed in substations and communicating with the central control system [i.e., supervisory control and data acquisition (SCADA)], or the computers in an office accessing the main servers, are connected through telecommunication networks.

Depending on the requirements of the connectivity, the network nodes location, the physical media used, the modulation of the telecommunication signals, the protocols used at different layers, and the general architecture, the resulting telecommunication network will be a different one. Different telecommunication networks may be better adapted to specific needs, both technically and economically.

2.1.2   Transport and Switching

A telecommunication network combines transport and switching technologies (Figure 2.1). Transport is the basic function present in any telecommunication network to carry the signal between two network nodes, transmitter and receiver. Switching is present when this information must go through one or several network nodes between the transmitter and the receiver, and different paths could be taken. These network nodes switch the information, selecting the next hop to reach the receiver, and can also provide some other optional functionality such as data prioritization and/or aggregation and disaggregation.

A comparison for the transport and the switching process is the road transportation. A car driving through a road connecting two cities is similar to the telecommunication transport process. When the car reaches a junction, the driver decides which direction should be taken to reach the final destination. This decision is similar to the switching process, with the only difference that it is normally made automatically by the telecommunication node.

There are more similarities between telecommunications and road transportation:

- There are speed limitations in the roads, as in the telecommunication channels.

- There are different categories of roads (regional, national, and so forth). In telecommunications, there are also different types of transmission channels.

- Several vehicles use the same roads simultaneously. In telecommunications several transmitters use the same transmission channel.

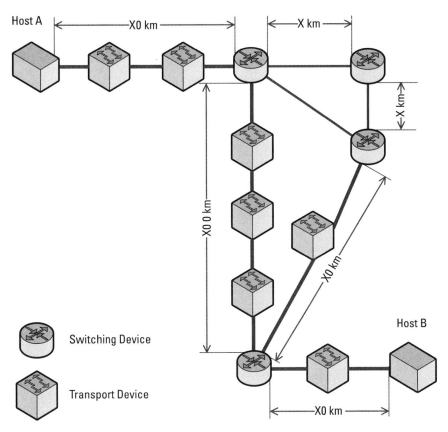

Figure 2.1 Transport and switching.

However, there are certain particularities specific to telecommunications:

- The switching mechanism is automatic. The system decides where to send the information based on its knowledge of the network and on the data included as part of the message (control data).

- Modern switching techniques allow information processing and not only switching. Thus, channel and transmission resources are used more efficiently (e.g., data compression and multiplexing).

2.1.3 Circuits Versus Packets

There are two different categories of switching technologies depending on the method used to choose the path between transmitter and receiver. They are circuit switching and packet switching.

Circuit switching works with fixed paths. With this method, prior to sending the information, transmitter, receiver, and intermediate devices need a connection to be set up (i.e., the circuit). It defines the whole path to be used for the information exchange.

This method is used, for example, in the classic telephone system. When a user makes a call, network nodes establish a physical circuit between the origin and the destination. This circuit is used during the whole conversation and is disconnected when the call is finished. The next time the user makes the same call, a new circuit will be established, probably through a different path, as the resources available in the network could be different each time.

Packet switching works with information divided in several pieces or blocks, called packets. There is no circuit set up from origin to destination prior to sending packets, and each packet is switched through the network to reach the destination. In order to switch a packet, intermediate devices need information included in the packet itself and in the device, which knows about the network and available resources. Thus, intermediate devices can automatically decide where to send each packet, in such a way that several packets from the same transmission can take different paths through the network. The receiver will have mechanisms to receive, order, and reassemble all the packets into the original information.

Packet switching is used, for example, in the Internet. When a user accesses a particular Web page, the information contained is divided into packets and sent from the server to the user's computer. The path for each packet may be different and will possibly arrive at different times, but eventually, thanks to the mechanisms defined in these technologies, the user will see the complete Web page in the browser.

It is also important to mention that there are some hybrid approaches which use both the circuit and the packet switching concepts. This is the case, for example, of Multiprotocol Label Switching (MPLS) (architecture specified in IETF RFC 3031). MPLS defines some preestablished paths for the different kinds of packets, which are tagged with a particular label. Every packet with the same label will be transmitted through its corresponding path or paths, if more than one are predefined.

2.1.4 Traffic

Traffic intensity is a measure of the amount of information being transmitted in a telecommunication network. The concept, again, is very similar to the one commonly used in road transportation.

One of the main challenges when designing a telecommunication system is the network dimensioning (i.e., the definition of the capacities of the network nodes and the links between them). These elements are typically limited and/or expensive resources, and a correct dimensioning is needed.

To determine the necessary resources to provide the required service, traffic engineering techniques are used. The main variable used in traffic engineering is the traffic intensity and its unit is the Erlang (E).

Traffic intensity has been defined in ITU-T Recommendation E.600 as "the instantaneous traffic intensity in a pool of resources is the number of busy resources at a given instant of time." One Erlang has been defined as "the traffic intensity in a pool of resources when just one of the resources is busy."

2.1.5 Multiple Access

In a telecommunication network where several devices are connected to the same transmission medium, it is necessary to define the mechanism to coordinate the activity of all devices sharing the network, avoiding simultaneous transmission that would limit or impede reception due to collisions. Multiple access mechanisms allow multiplexing (combining and transmitting together in an orderly manner) several signals over the same channel.

Multiple access may be achieved with the following mechanisms (Figure 2.2):

- Frequency division multiple access (FDMA): Transmission is done in different frequency slots or channels, known to every device in the network. Devices will use a particular frequency, assigned to them with particular modulations. An example of this technique is implemented in the radio frequency bands, and it is how telephone copper pairs are shared for voice and data transmission in digital subscriber line (DSL) technology.

- Time division multiple access (TDMA): The channel is divided in time slots known to every device in the network. Time-slot duration and assignment can be fixed or variable. Devices will use a particular time slot,

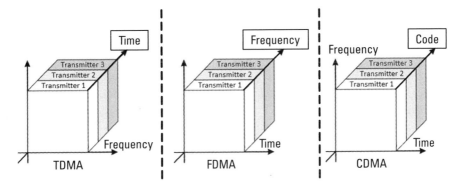

Figure 2.2 Multiple access techniques.

assigned to them. Well-known examples of systems using this technology are time division multiplexing (TDM) transmission technologies such as plesiochronous digital hierarchy (PDH).

• Code division multiple access (CDMA): The channel is divided assigning a particular code to each device of the network. The devices will use it to encode the signal transmitted and thus, the receiver can decode the signal with the inverse operation using the same code in reception. The transmission is usually based on spread-spectrum techniques. An example of this technique is implemented in third generation (3G) mobile radio technologies such as Universal Mobile Telecommunications System (UMTS).

2.1.6 Modulation

Modulation has been defined by the ITU as "a process by which a quantity which characterizes an oscillation or wave follows the variations of a signal or of another oscillation or wave." In simpler words, it is a process by which the information to be sent (analog or digital) is used to modify one or more characteristics of the carrier signal transmitted.

The modulation process adapts the signal to the optimal conditions in each particular medium, considering aspects such as distance to be covered, bandwidth, and robustness. Normally, gaining robustness means losing bandwidth and vice versa.

Classic examples of modulations are analog amplitude modulation (AM) and frequency modulation (FM). In AM the amplitude of the carrier signal is modified following the information to be sent; in FM, the parameter that varies is the signal frequency. Modern systems use digital modulations, usually derived from their analog equivalents.

2.1.7 Signal, Noise, and Interference

The transmitted signal is usually called the desired signal in telecommunication networks; the optimal scenario is when the signal received matches exactly the signal sent. However, there are three main effects (Figure 2.3) affecting the quality of the reception: propagation, interference, and noise.

Propagation effects are those that modify the signal within the media due to their physical constraints:

• Attenuation: It is the loss of signal power. It is normally a function of distance.
• Distortion: It is the change in signal shape.

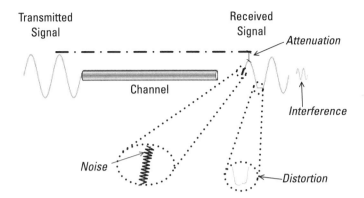

Figure 2.3 Signal propagation.

Interference is the combination of telecommunication signals generated by the receiver and/or undesired transmitters affecting the correct reception of signals. These signals reduce the receiver's capability to detect the desired signal.

Noise is any disturbance (including in some cases interferences) that hinders or makes impossible the reception of the signal by the receiver. It is always present in telecommunication networks and it is one of the main problems to be solved by means of appropriate modulation and coding techniques.

Signal-to-noise ratio (SNR) is a widely used parameter in telecommunication network design to estimate the performance of receivers. It has been defined by the IEC as "the ratio of the wanted signal level to the electromagnetic noise level as measured under specified conditions." It measures the difference between the power of the desired signal and the power of the noise. SNR should be as high as possible in order to facilitate the reception of the signal.

2.1.8 Bandwidth

Bandwidth has been defined by ITU-R Recommendation SM.328 as "the width of the band of frequencies occupied by one signal, or a number of multiplexed signals, which is intended to be conveyed by a line or a radio transmission system."

Apart from this formal definition, there is another commonly accepted way of referring to bandwidth when digital information is involved. It takes into account the capacity of the telecommunication system, and expresses bandwidth as the amount of information that a system is able to manage in bits per second (bps). This data rate is directly dependent on the frequency bandwidth, the modulation, and the coding process.

2.1.9   Radio-Related Concepts

2.1.9.1   Electromagnetic Spectrum

The IEC has defined frequency spectrum as "the range of frequencies of oscillations or electromagnetic waves which can be used for the transmission of information." Although this is a generic definition applicable to any transmission medium, it is generally accepted in telecommunications that spectrum refers to the radio spectrum used for wireless transmissions (Figure 2.4). Table 2.1 contains the list of frequency bands in which the radio spectrum is divided as stated by the ITU RR (2012 Edition, Article 2, Section I) [1]. Each particular band has its own name and characteristics (propagation conditions, capacity, bandwidth, and so forth).

The microwave concept needs also to be mentioned as it is extensively used in industry. Microwave has been defined by the IEC as "radio wave with a wavelength sufficiently low to allow use of technologies such as waveguides, cavities, or planar transmission lines; the corresponding frequencies are higher than about 1 GHz." In free space, the wavelength can be calculated as the speed of light ($3 \cdot 10^8$ m/s) divided by the frequency of the radio wave, expressed in hertz (see ITU-T Recommendation K.52).

2.1.9.2   Antenna (Radiating System)

An antenna is a transducer that allows the adaptation of the signal to be transmitted to a wireless medium. It is a set of metallic elements arranged in a way to adapt the signal to be transmitted and received through the air. The design is particular for each frequency band.

The main characteristics of antennas are:

- Gain: Although not a formal definition, antenna gain indicates the ability of an antenna to radiate the injected power in a certain direction (the

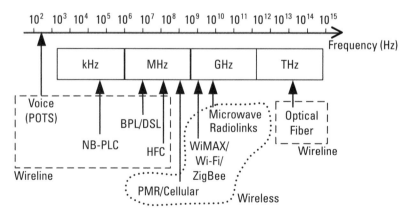

Figure 2.4   Electromagnetic spectrum.

Table 2.1

Radio Spectrum Bands

Band Number	Symbol	Frequency Range (Lower Limit Exclusive, Upper Limit Inclusive)	Corresponding Metric Subdivision
4	VLF	3 to 30 kHz	Myriametric waves
5	LF	30 to 300 kHz	Kilometric waves
6	MF	300 to 3,000 kHz	Hectometric waves
7	HF	3 to 30 MHz	Decametric waves
8	VHF	30 to 300 MHz	Metric waves
9	UHF	300 to 3,000 MHz	Decimetric waves
10	SHF	3 to 30 GHz	Centimetric waves
11	EHF	30 to 300 GHz	Millimetric waves
12		300 to 3,000 GHz	Decimillimetric waves

Note: Band N (N = band number) extends from 0.3×10^N Hz to 3×10^N Hz

same concept is applicable in reception). It usually refers to the direction of maximum radiation.

• Beamwidth: It is related to the antenna gain. It is the solid angle at which this gain does not fall below a certain value (–3 dB gain is usually considered).

2.1.9.3 Propagation Phenomena

Apart from the generic propagation effects explained in Section 2.1.7, there are other phenomena affecting the propagation of signals through the air (Figure 2.5):

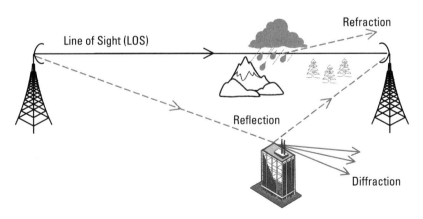

Figure 2.5 Radio propagation effects.

- Reflection: When the radio signal impacts with an object, the signal is reflected with different characteristics depending on the surface and the wave itself.

- Diffraction: When the signal finds a corner, multiple instances of the same signal are generated, traveling in all directions. If the size of the obstacle is comparable to the wavelength, we refer to it as scattering.

- Refraction: The speed and the direction of the wave changes when it goes through a nonhomogeneous medium.

- Transmission: This effect allows a signal to partially go through objects (e.g., nonmetallic surfaces).

2.2   Digital Telecommunications

Although telecommunications cover both analog and digital telecommunications, this book is focused on digital telecommunications, as the best option to provide modern smart grid services.

In an analog world, we refer to digital telecommunications because the two end-points of a telecommunication channel make a digital interpretation of the transmitted signals (Figure 2.6).

The main advantages of digital telecommunications are:

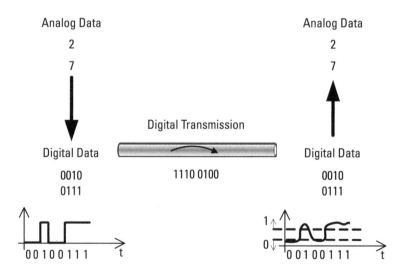

Figure 2.6   Digital telecommunications.

- Improved reception quality to overcome problems present in analog technologies (e.g., noise accumulation, solved with signal regeneration);
- Increased efficiency of telecommunication channels use (i.e., higher data rates), thanks to the implementation of multilevel digital modulations (one level or symbol represents more than 1 bit);
- Error correction signal processing capability.

2.2.1 Open Systems Interconnection Reference Model

A digital telecommunication network provides data transfer services between two digital end-points (or hosts). This transport service is used by the applications at each end-point to reach the remote host. For example, the SCADA system running in a central server sends a request to an IED or RTU in a substation. The SCADA operates at the application level and manages to communicate with the remote end-point, transparently to the underlying low-level mechanisms of the telecommunication network.

This abstraction is achieved with the open system interconnection (OSI) reference model. The information managed at application level traverses several layers that, one by one, accommodate the information to be sent. In order for the received information to be understood, the process should be standard at both end-points. For this purpose, specific protocols are defined for every layer of the model. A formal definition of this layered architecture has been proposed in ISO/IEC 7498-1:1994, *Information technology—Open Systems Interconnection—Basic Reference Model: The Basic Model*, also adopted by the ITU-T in Recommendation X.200 (Figure 2.7).

The OSI reference model defines seven layers. Each layer provides service to the layer above it. Application, presentation, session, and transport layers are called the upper or host layers and they operate end-to-end. Network, data link, and physical layers are called the lower or media layers and operate hop-to-hop [2]. In other words, intermediate nodes on a telecommunication path only apply the media layers while the end-point nodes (origin and destination) apply both the media and the host layers. From the telecommunications perspective, lower layers together with the transport layer are usually considered as telecommunication-specific. The rest of the upper layers are application and data-formatting layers with no direct impact in the telecommunication network itself. An example of this separation is found in the ICT acronym: the *I* is related to the three upper layers, while the *C* is related to the four telecommunication-specific layers.

Each OSI model layer defines a protocol or a set of protocols to specify its functions. The most common functions defined for every protocol are:

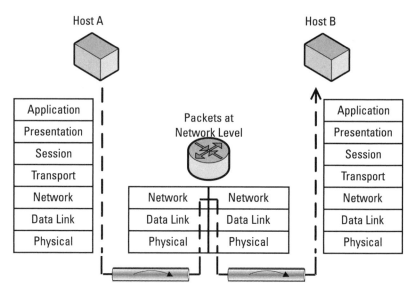

Figure 2.7 OSI reference model.

- Packet data unit (PDU) formats, with header, payload, and checksum;
- Control PDUs;
- Segmentation and reassembly procedures for upper-layer information;
- Acknowledgment procedures;
- Error control and recovery procedures.

2.2.2 Physical Layer

The lowest layer of the OSI reference model, commonly known as the PHY layer, defines the physical characteristics of the telecommunication channel and the way it is used.

As defined by ISO/IEC, "the Physical Layer provides the mechanical, electrical, functional and procedural means to activate, maintain, and de-activate physical-connections for bit transmission between data-link-entities. A physical-connection may involve intermediate open systems, each relaying bit transmission within the Physical Layer. Physical Layer entities are interconnected by means of a physical medium."

There are different physical media that can be used in telecommunications. Radio (wireless) uses the air as the "physical" medium; wireline telecommunications use different types of cables (e.g., optical fiber, coaxial, twisted pairs). Depending on the physical media, the telecommunication channels show different properties:

- Propagation speed: It is the physical velocity based on the propagation of light. In free-space (air) it is 300,000 km/s, but lowers inside other media.

- Transmission rate: It is related to the modulation aspects of the transmission, and it can be expressed in bits per second (e.g., Ethernet typically goes from 10 Mbps to more than 1 Gbps).

Once the physical medium is defined, the main characteristics of the telecommunication transmission should also be specified:

- Frequency: It defines the central frequency or frequencies of the transmission.

- Bandwidth: Once the channel is defined, the protocol should also define the bandwidth (see Section 2.1.8) of the signal. It is usually much lower than the frequency used. With F as the frequency and B as the bandwidth, the occupation goes from F – B/2 to F + B/2.

- Modulation: As covered in Section 2.1.6, the modulation defines the way that frequency and bandwidth are used.

- Data rate: It is the consequence of the previous parameters over a certain medium.

With all these parameters, a PHY layer protocol defines the way the information is transformed in bits and how these bits are sent to the end-point. The remote host, at this level, will decode the information, unless any impairment (see Section 2.1.7) corrupts the received signal to a level that makes it unintelligible.

2.2.3 Data Link Layer

The rules for the different devices in the network to access the transmission medium and use the network resources in an orderly manner are defined in the data link layer (DLL).

As defined by the ISO/IEC, "the Data Link Layer provides functional and procedural means for connectionless-mode among network-entities, and for connection-mode for the establishment, maintenance, and release data-link-connections among network-entities and for the transfer of data-link-service-data-units. A data-link-connection is built upon one or several physical-connections."

The basic functions of the DLL can be divided in two groups: multiple access and resource sharing, and traffic control.

Multiple access functions (FDMA, TDMA, CDMA; see Section 2.1.5) manage the assignment of the channel resources to the different network devices. Resource sharing is implemented by each technology with different procedures. As an example, the token concept could be applied. With this approach, the device holding the token has the permission to transmit. When the transmission is finished, it passes the token to the next device in the list, which is the next one allowed to transmit. Devices can only transmit when they hold the token and must remain idle or in reception mode the rest of the time.

A particular solution to the media access control is carrier-sense multiple access (CSMA). In the protocols implementing this solution, each device, prior to sending a packet to the channel, senses the channel to check whether it is occupied by another ongoing transmission. If there is no device transmitting, it will take the decision to transmit. If there is other device occupying the channel, it needs to wait until the channel is free again. There are two main variants of CSMA protocol:

- CSMA with Collision Avoidance (CSMA/CA): If the channel is occupied, the device waits a random time interval to retry.
- CSMA with Collision Detection (CSMA/CD): When a node is transmitting, it also keeps on detecting if there is a collision. If it detects a collision, the device stops the transmission immediately in an attempt to optimize the use of the channel.

Traffic control functions are implemented in the DLL. Traffic scheduling and admission control are important for the implementation of quality of service (QoS). Some other functions define how downlink and uplink (transmission and reception links) are controlled. The most common options are half-duplex, when only one of the end-points can be transmitting at a given time, and full-duplex, when both end-points can be transmitting simultaneously.

2.2.4   Network Layer

The network layer defines addressing and packet format. The information to be transmitted is divided into blocks, which are passed to the data link layer together with information related to the network layer, valid for the receiver to understand the message.

As defined by ISO/IEC, "the Network Layer provides the functional and procedural means for connectionless-mode or connection mode transmission among transport-entities and, therefore, provides to the transport-entities independence of routing and relay considerations."

A unit of transmission is usually called a PDU and is exchanged by the same entities at both ends of the telecommunication channel. Every PDU consists of the following elements (Figure 2.8):

- Header: It is the information added by the corresponding network layer protocol. It is used by the packet to reach the remote destination in the conditions needed by the transport layer, and by the remote element to understand the PDU.

- Payload: It is the information itself, coming from the upper layer.

- Checksum: It is the group of bytes used to verify the integrity of the received packet, to avoid transmission errors or external modifications. It is generated applying mathematical operations to the bytes of the PDU or part of it (e.g., the header). Cyclic redundancy check (CRC) is a very common checksum.

2.2.5 Transport Layer

The transport layer provides data transfer from source to destination, managing data flow and QoS. The transport layer lies in the separation between host layers and network layers, and it is the last layer, which can be strictly considered a telecommunication layer.

As defined by the ISO/IEC, "transport-service provides transparent transfer of data between session-entities and relieves them from any concern with the detailed way in which reliable and cost effective transfer of data is achieved."

Transport layer implementations can be classified as reliable or unreliable. Reliable transport is achieved using connection-oriented protocols that implement mechanisms to establish, maintain, and terminate the connection in an orderly manner. These mechanisms include sequence numbering for PDU, acknowledgment packets to confirm or not confirm reception, and data loss recovery through retransmissions. On the contrary, unreliable transport, or transport without the ability to detect duplicated or lost packets, is provided by nonconnection-oriented protocols. These protocols do not implement mechanisms to establish or manage connections.

Transmission Control Protocol (TCP) is an example of reliable transport. User Datagram Protocol (UDP) is an example of unreliable transport. It is important to highlight the difference between TCP and UDP for the design of any telecommunication network. If UDP is used, there is no retransmission

Header	Payload	Checksum

Figure 2.8 Network layer packet format example.

mechanism at this layer; the end-to-end packet loss could be high in nonreliable and/or shared physical media. However, if the telecommunication channel is highly reliable (e.g., wireline solutions), UDP can reduce the amount of bulk network traffic.

2.2.6 TCP/IP and the OSI Model

The OSI reference model is an internationally accepted model for the layered architecture of digital telecommunication networks. The model is applied in different ways depending on the technology, but it is not always strictly followed. Many implementations adjust the model for the sake of efficiency.

A clear example of this is the TCP/IP protocol stack [3], used worldwide. TCP/IP architecture is based on 4 layers instead of 7 (Figure 2.9).

TCP/IP includes two different protocols, TCP and IP, fitting in the transport layer and the network layer of the OSI model, respectively. Their main characteristics (scalability and flexibility), together with the great success of their use in the Internet, have turned TCP/IP into the most widely accepted protocol stack. Many applications have been developed over TCP/IP and other applications have been adapted to be used over it. IP will be further covered in Chapter 5 as one of the key technologies for smart grids.

The TCP/IP stack offers:

- Network layer: It matches both the physical and the data link layer of the OSI reference model. It defines the physical media together with the data link characteristics, for example, Ethernet over multipair copper cables, as defined in IEEE 802.3.

- Internet layer: It matches the network layer of the OSI reference model. This layer is originally defined in IETF RFC 791, the specification of IP.

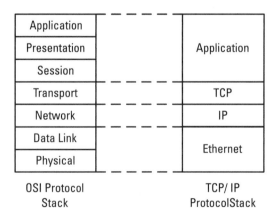

Figure 2.9   TCP/IP protocol stack versus the OSI reference model.

- Transport layer: It matches the transport layer of the OSI reference model. This layer is originally defined in IETF RFC 793, the specification of TCP.

- Application layer: It matches the three upper layers of the OSI reference model.

2.3 Physical Media

The physical layer is a key element for the smart grid. It is the element that has to do with real infrastructure, assets, rights of way, and rights of use, and it can become a bottleneck in any telecommunication network implementation when it is not available.

The first important decision to be made, when defining a telecommunication network, is the transmission media to be used in each network segment and location. This decision will condition the rest of the system. The physical media selection eventually brings further alternatives to decide on, as each physical medium is different from the rest.

2.3.1 Radio

Radio (wireless) is the supporting medium making the transmission of data in the form of electromagnetic waves over the air (Figure 2.5). The main advantage of this physical layer is that it is already present everywhere, without the need of any previous cable deployment.

The concept and the basic physical phenomena allowing the existence of radio technologies are the same for any frequency range. However, some characteristics are specific of each frequency band. Lower-frequency bands have better propagation conditions but lower bandwidth and, consequently, lower capacity. On the contrary, higher frequencies have poorer propagation characteristics but higher bandwidth and higher capacity. Propagation characteristics are one of the inputs (apart from economics, regulation, and so forth) that should guide the decision on the frequency band to be used based on the area to cover, the nature of the assets where the services will be delivered and the capacity [e.g., the advantage of microwave frequencies is that the available bandwidth is higher than with very high frequency (VHF) and ultrahigh frequency (UHF); the main disadvantage is that propagation is more difficult, as direct line of sight (LOS) is generally required].

Radio technologies have been widely used since they were first made available at the end of the nineteenth century. They are present in different applications from commercial radio and TV broadcasting to cellular systems and remote control. The evolution of radio telecommunications has been determined by the constant exploration of higher-frequency uses, available as the technology progressed and made faster and lower-cost electronics, by improvements in transmission and reception techniques to be able to decode weaker signals even in the presence of noise and interference, and by the search of global applications that could be used anywhere globally, avoiding the costs of wireline infrastructure.

The ITU-R is the global reference for radio frequency international coordination. Although national governments finally review the frequency bands to be accessed for each spectrum user, the ITU establishes the global framework (e.g., satellite or ionospheric telecommunications). The ITU RR are the fundamental reference of this coordination.

2.3.2   Metallic Telecommunication Cables

Metallic cables are in the origin of the first wireline transmission in telegraphy. The original use in telegraphy for the transmission of simple data signals (e.g., Morse code) evolved into more complex analog voice transmission through copper pairs after the telephone invention. Eventually, copper pairs were (again) reused to transmit data with the advent of xDSL techniques.

The basic metallic components of the transmission medium are often grouped in a cabled structure that gathers a number of conductors and gives them a higher mechanical robustness. There are two main types of metallic cables widely used in telecommunications (Figure 2.10):

- Multipair copper cables (twisted or untwisted): These are one of the most popular media in telecommunication networks, due to their relative low cost and simple installation. They are normally used for short- and medium-length connections (hundreds of meters or even a few kilometers), as well as for voice telephony services. Crosstalk (interference among pairs) is a side effect of the grouping of conductors in the same structure.

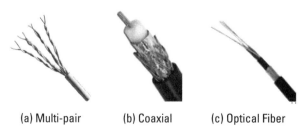

(a) Multi-pair   (b) Coaxial   (c) Optical Fiber

Figure 2.10   Cables example.

• Coaxial cables: Even if they are more recent and less popular than the multipair cable, their transmission characteristics are better. Their structure is that of a central metallic wire, surrounded by a metallic screen fitted over a dielectric material that separates both conductors. Their higher costs and installation difficulty have limited their use to some particular applications such as cable television and broadband short-range data transmission.

In the case of multipair copper cables, the number of pairs together in a cable varies greatly, from very few in the last mile to hundreds of them in other network segments. However, in the case of coaxial cables, it is not common to find groups of them in a single structure; on the contrary, it is usual to find different diameters of the coaxial structure and the cable design and robustness applied to this single unit.

Regarding the basic types of conductors, multipair copper cables are usually classified according to their diameter. It is common to find the reference to American Wire Gauge (AWG). Coaxial cables typically follow the radio guide (RG) denomination (e.g., RG-11, RG-6, and RG-59).

Electrical power lines can be also considered in this section. However, power line communications (PLC) will be covered in Section 2.3.4 due to its relevance for smart grids.

Cabling standards are necessary to define a uniform set of infrastructure elements that, together with the installation instructions and corresponding tests, can guarantee performance. For example, ISO/IEC 11801, EN 50173, and ANSI/TIA/EIA 568-B are widely used in the case of buildings. Categories are defined to achieve a certain system performance expressed in terms of data rate and applications supported. The MICE concept was developed to refer to environmental conditions the cables themselves need to be prepared to support: M = mechanical rating, I = ingress rating, C = climatic/chemical rating, and E = electromagnetic rating.

Regarding cable structures construction, cables have historically been used to cover long distances over any type of conduction (e.g., suspended on poles, buried with or without ducts). This has brought important developments in cable designs, the way the cables need to be assembled and the materials to be used to adapt them to the different installation constraints and achieve a lifespan aligned with the high costs of the networks they help to configure.

2.3.3 Optical Fiber

Optical fiber-based telecommunications have been a quantitative breakthrough in bandwidth and range extension, based on the transmission of data as light pulses through plastic or glass cores.

Compared to other transmission media, fiber optics is a modern one, as laser was invented in the 1960s. From 1975 when the research phase on fiber optics telecommunication systems started, its evolution can be divided into five stages (see ITU-T Rec. G.Sup42, *Guide on the use of the ITU-T Recommendations related to optical fibres and systems technology*):

- The first phase of fiber optics systems operated near the wavelength of 850 nm, which is now called the first window. These systems were commercially available by 1980 with multimode fibers [typically used in local area networks (LANs)]. Their reach extended up to 10 km with a transmission speed of around 40 Mbps. ITU-T Recommendations G.651 and G.956 (now G.955) standardized them.

- The second evolution of fiber optics systems operated near 1,300 nm, that is, the second window. These systems were commercially available by the early 1980s, initially limited to bit rates below 100 Mbps due to the dispersion in multimode fibers. With the development of single-mode fibers (extensively found now in long-range, tens of kilometers, applications) these systems reached distances of around 50 km and 1.7 Gbps data rates by 1988. Among others, ITU-T Recommendations G.652, G. 957 and G.955 standardized them.

- The third evolution of fiber optics systems operated near the 1,550 nm wavelength, where the attenuation becomes minimal at values of around 0.2 dB/km. This wavelength is referred to as the third window. These systems were commercially available by 1992 with dispersion-shifted fibers. They allowed for distances of around 80 km and 10 Gbps data rates. Among others, ITU-T Recommendations G.653, G. 957, G.955, and G.794 standardized them.

- The fourth evolution of fiber optics systems operated also near the 1,550 nm wavelength making use of optical amplifiers to increase distances and of wavelength division multiplexing (WDM) techniques to multiply the data rate (WDM is conceptually similar to FDMA multiplexing). By 1996, distances of 11,600 km were reached with 5 Gbps data rates for intercontinental applications (obviously with optical amplifiers). Among others, ITU-T Recommendations G.655, G.694, and G.695 standardized them.

- Currently, fiber optics systems are still evolving:
  - Increasing the capacity of WDM systems by means of transmitting additional channels, with more capacity per channel, and reduced channel spacing;

- Reducing optical/electrical/optical conversion needs, using optical devices in a so-called active optical network (AON);
- Reducing the cost and size of optical/electrical/optical conversions;
- Developing passive optical networks (PONs) for the access network in Fiber to the X (FTTx) architectures.

Optical fiber cables (Figure 2.10) are the only media that guarantee a continuous expansion of capacity and range. This makes optical fiber cables a very suitable telecommunication medium for any purpose, specifically for smart grid. This is so not only because of the high bandwidth provided and the long-distance reach, but due to the fact that the fiber itself is a nonconductive material. This is very useful when the transmission medium needs to be integrated in high-voltage (HV) premises with electromagnetic compatibility (EMC) constraints that happen to create disturbances in conductive metallic media. As optical fiber is a dielectric material, the telecommunication network is fully insulated and easily complies with safety considerations.

2.3.4 Power Lines

Power lines are essentially metallic conductors that are used for the very special 50- or 60-Hz transmission of electricity. In the same way that copper pairs can be used both and in parallel for voice and data transmission, power lines can be used for electricity and also for telecommunication data transmission; this is the basic concept behind PLC (Figure 2.11). Different frequencies, modulations, and materials make channel conditions vary in the different power lines and PLC systems.

From the physical layer perspective, PLC can be classified as follows [4]:

- Ultranarrowband PLC: The physical layer of these systems works in ultralow frequency (0.3–3 kHz) or super low frequency (30–300 Hz), reaching tens and even hundreds of kilometers with very low data rates of around 100 bps or less.

- Narrowband PLC: The physical layer works in frequencies from 3 to 500 kHz, including CENELEC A-band (3–148.5 kHz), the FCC band (10–490 kHz), and the ARIB band (10–450 kHz). They reach distances from hundreds of meters to some kilometers, depending on the power line type. From the data rate point of view, we can find two classes:

  - Low data rate: Single carrier technologies providing a few kilobits per second;

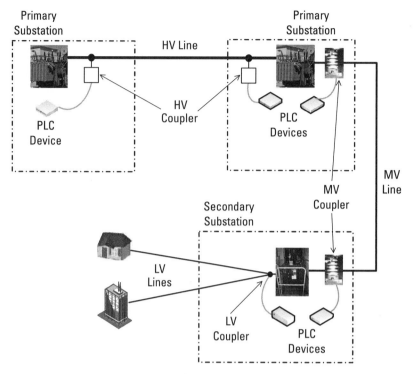

Figure 2.11   Power line communications (PLC).

- High data rate: Multicarrier technologies providing hundreds of kilobits per second.
- Broadband PLC or broadband power line (BPL): The physical layer works in frequencies from 1.8 to 250 MHz, reaching distances of hundreds of meters and perhaps kilometers, depending on the frequency and the power line cables. Data rates are up to hundreds of megabits per second.

The historical evolution of PLC, used in electrical companies for over a century, has been driven by the applications needed by the industry itself. PLC technologies' origin cannot be precisely identified, but there are patents on the concept by the early twentieth century. The first widely used PLC application at the end of 1920s was voice telephony over HV power lines to reach long distances, wherever telephone lines were not available. There are also early evidences of PLC applications for meter reading (early twentieth century), and it was in the 1940s and 1950s when several utility companies started developing PLC systems to control the load in the grid. These systems can be considered as the seed of current PLC systems used for smart grids.

In terms of standardization, the market has historically been driven by vendor-specific implementations and proprietary systems. In the twenty-first century, the situation has changed to some extent and several standards have been published. In Europe, as a consequence of mandates M/441 and M/490, smart metering systems based on high date rate narrowband PLC, have managed to standardize the PLC technologies they use as IEEE 1901.2 and ITU-T Recommendations G.990x.

With regard to BPL, the Open PLC European Research Alliance (OPERA) [5] project focused on the use of BPL for the access segment [medium-voltage (MV) and low-voltage (LV) grid]. OPERA demonstrated the technical feasibility of BPL Internet access (competing with xDSL and HFC technologies), and the possibilities to use MV and LV grids as physical media for smart grid telecommunications. Eventually, BPL technology was standardized as IEEE 1901 and ITU-T Recommendations G.9960/G.9961.

References

[1]   International Telecommunication Union-Radiocommunication (ITU-R), *Radio Regulations Edition 2012*, Geneva, 2012.

[2]   Penttinen, J. T. J., (ed.), *The Telecommunications Handbook: Engineering Guidelines for Fixed, Mobile and Satellite Systems*, New York: Wiley, 2015.

[3]   Puzmanova, R., *Routing and Switching, Time of Convergence?*, Boston, MA: Addison-Wesley Professional, 2002.

[4]   Galli, S., A. Scaglione, and Z. Wang, "For the Grid and Through the Grid: The Role of Power Line Communications in the Smart Grid," *Proceedings of the IEEE*, Vol. 99, No. 6, pp. 998–1027.

[5]   Open PLC European Research Alliance (OPERA) for New Generation PLC Integrated Network, "Project Details," 2008. Accessed March 21, 2016. http://cordis.europa.eu/project/rcn/71133_en.html; http://cordis.europa.eu/project/rcn/80500_en.html.

3

Electric Power System Concepts for Telecommunication Engineers

Readers will usually have a stronger background in either telecommunications or power systems. This chapter will give a general view of useful topics in the area of power systems and electric grid operations. The electric power system is defined by the International Electrotechnical Commission (IEC) as "all installations and plant provided for the purpose of generating, transmitting and distributing electricity."

3.1 Fundamental Concepts of Electricity

The American National Academy of Engineering [1] has referred to electricity as "the workhorse of the modern world" as it is critical to the daily lives of most of the world's population. Most of the major engineering achievements that we enjoy today would not have been possible without the widespread electrification that occurred in the twentieth century [2].

3.1.1 Electric Charge and Current

Electric charge, or just "charge," is a fundamental property of all matter, and electricity treatises usually begin by assuming its existence. Charge can be defined in simple terms as the property that causes particles to experience a force when placed near other similar ("charged") particles. This force has historically been characterized and measured as a field (force per unit of charge) and more specifically as an electromagnetic field, which is considered to be one of the four fundamental interactions of nature (the others being weak interaction, strong interaction, and gravitation [3]). The electromagnetic field, physically produced by charged particles, affects the behavior of other charged particles (exercising a force).

Charge is generally understood to be quantized, that is, there is a specific amount of charge that is considered the smallest unit: the elementary charge or charge carried by a proton, equal to 1.602×10^{-19} coulombs. Charge and, in general, electromagnetic force shows either an attractive or repulsive behavior (unlike gravity, which is always attractive). By convention, the concepts of "positive" and "negative" charges are used, meaning that charged particles of the same sign repel each other, while charged particles of different signs attract each other.

The study of charges that are not moving is referred to as electrostatics, although it is not what we find in electric power systems; in practice, power systems are always based on the movement of charges through conductors. For any given point in space, the amount of charge that moves through that point per unit of time is defined as electric current. Current is a measure of the flow of charge expressed in amperes, so that 1 coulomb per second equals 1 ampere. As an analogy that is usually made, current would be equivalent in electrical terms to the rate of water flow through a pipe, which in the electricity world would represent a cable or power line.

3.1.2 Voltage

At a given instant in time, a local concentration of charges will have a tendency to spread out if they are of the same sign. Charges will be pressed to move in the same way that, following with the analogy, pressure difference forces water in a pipe. More specifically, the concept indicates the potential energy of each of the charges, understood in a similar way to the potential energy of an object at a certain elevation, such that it represents the work that needed to be done to locate the object at the elevation or, in our case, the work to place the charges in their current position, overcoming the attractive and repulsive forces of other charges.

Charge depends exclusively on the particle itself, while the potential energy depends on a charge's location with respect to another reference location. This potential is formally known as electric potential difference, electric potential, or electric tension, but in common language it is referred to as voltage.

A more rigorous definition of voltage would be to describe it as electric potential energy per unit charge. It thus means that the electric potential is a per-unit charge quantity, not to be confused with the electric potential energy itself.

Similarly to the previous analogy of potential energy of an elevated object, the reference of electric potential can be chosen at any location, so the difference in voltage is the quantity that is physically meaningful. The voltage difference U_{ab} between point a (represented by position vector r_a, a mathematical construct defined by an amplitude and an angle with respect to a reference point) and point b (represented by position vector r_b) is the work needed to be done, per unit charge, against the electric field E to move the charge from point a to point b. Mathematically, this is expressed as the line integral of the electric field along that path.

$$U_{ab} = \int_{r_a}^{r_b} E \cdot dr$$

Voltage is measured in volts; 1 volt is 1 newton per coulomb (electric field) multiplied by 1 meter (distance). Different sources, some of them also focused on Smart Grids [4, 5], develop the basics of these concepts.

3.1.3 Direct Current and Alternating Current Systems

When, in the simple circuit in Figure 3.1, voltage is a constant value V and current is unidirectional also with a constant value I, the system is usually referred to as direct current (dc).

When the voltage source actually changes its V value with time, usually the system is referred to as alternating current (ac) and described as $v(t)$. In power systems the variation with time usually follows a sine wave shape:

$$v(t) = \sqrt{2}V \sin\left(2\pi f_s t\right)$$

where f_s represents the system frequency (called the line frequency or the power frequency in electric grids). It is obvious that, for the sine shape voltage source defined above, the voltage level at the terminals changes from a maximum of $\sqrt{2}V$ volts to a minimum of $-\sqrt{2}V$ volts. This practically means that the terminal with the higher potential will alternatively be one or the other. Thus, the charges will find themselves flowing in one direction half of the time and in the

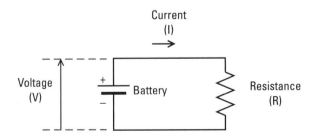

Figure 3.1 Simple electric dc circuit.

opposite direction the rest of the time. Positive and negative terminals, in the sense that they could be defined for dc, make no more sense; now it could be said that the polarity of the voltage source changes over one full period of the sinusoid.

Voltages and current are both represented by sinusoids. The parameters typically associated with sines can be used for ac circuit characterization (see Figure 3.2).

The maximum value of the waveform is called the amplitude. Both voltage amplitudes and current amplitudes will be found in power systems. The amplitude of the ac voltage in a standard 230V outlet is 325V. The 230V always refers, by convention, to the root mean square (rms) value of the voltage (it represents the equivalent dc voltage value that would perform the same amount of work). In the case of perfect sinusoids, the rms value is equal to the amplitude value divided by a $\sqrt{2}$ factor.

Frequency describes the rate at which the current/voltage waveforms oscillate. It is measured in number of cycles per unit of time, its unit being the hertz that represents the inverse of a second. In large power systems the frequency is a fundamental parameter that has to be kept accurate and stable. Two normalized values are used worldwide: 50 Hz and 60 Hz.

The period of time that it takes for an ac waveform to perform one full cycle and return to its original value with its original slope is the inverse of the

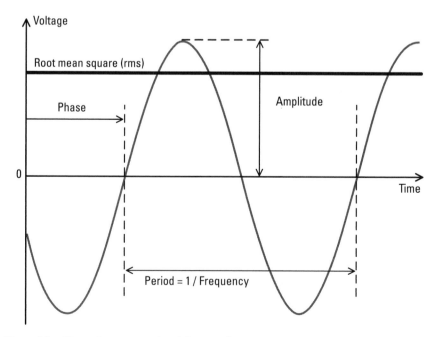

Figure 3.2   Alternating current electricity waveform.

frequency value and it is simply called the period. The phase angle (or simply "phase") of a sinusoid is a measure of when the waveform crosses zero relative to a previously established time reference. Phase is expressed as a fraction of the ac cycle and is measured in radians (ranging from $-\pi$ to $+\pi$).

Most power systems today are ac, although dc power is also employed in some parts of the world for transmission at very high voltages over long distances [6]; recently, there has been renewed interest and debate on the use of dc for large-scale transmission grids.

3.1.4 Impedance Elements

Impedance is the parameter of the loads that fixes the relationship between voltages and currents.

3.1.4.1 Phasors

In the conditions set forth up to now, the frequency of all voltage and current waveforms is the same, and the amplitudes are directly related to the rms values by a $\sqrt{2}$ factor, so each sinusoid is uniquely explained by its rms value and its phase angle. Thus, we can represent these sinusoids in the form of phasors (i.e., phase vectors), which are vectors that can be more easily used than the sinusoids they represent (Figure 3.3).

A voltage phasor **V** or a current phasor **I** is represented as a vector with its magnitude equal to the rms values of V and I and an angle equal to 0 and $-\varphi$, respectively.

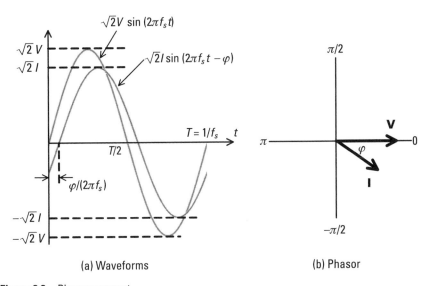

(a) Waveforms (b) Phasor

Figure 3.3 Phasor concept.

3.1.4.2 Resistance

In a dc circuit, resistance is the magnitude that relates voltage and current. As voltage represents the pressure difference between charges, and current is the rate of flow of those charges, it immediate follows that a certain voltage causes a proportional current to flow. While flowing, charges will encounter obstacles in their path. For a given voltage, a higher resistance will mean that the flow of current is low, while a lower resistance implies a larger flow of current. This is expressed as Ohm's law, which is fundamental for circuit analysis.

$$I = \frac{V}{R}$$

In a simple circuit (Figure 3.1) where there is a voltage source (battery) of the constant value V (measured in volts) connected to a resistance of value R (measured in ohms, with 1 ohm being 1 volt per ampere), there will be a current of I amperes flowing from the voltage source terminal with the higher electric potential (positive terminal) to the one with the lower electric potential (negative terminal). In the case of a resistance with infinite value, there would be no current (this is called an open circuit) and in the case of an ideal resistance with zero value, there would be an infinite flow of current for a nonzero V value (this is called a short circuit).

In the case of power grid design, materials with high-resistance properties (i.e., insulators) become fundamental for the target of obtaining low conductivities and keeping current flows safely contained. Ceramics and plastics are used extensively as insulating materials.

3.1.4.3 Reactance

The resistance value R shown up to now has been the representation of an ideal ohmic element (a resistor) with a real and constant value. This allows for in-phase voltage and current sinusoids, which is usually not the general situation of power systems.

Any flow of charges due to an existing electric field actually generates a different, complementary field called magnetic, so more intense currents imply more intense magnetic fields too. Changing currents (the case in ac systems) generate varying magnetic fields, and it is this variation of electric current that in turn induces an electric voltage opposite to the changing current. Induced voltages are proportional to the rate of change (i.e., first derivative) of currents.

$$v(t) = L \frac{di(t)}{dt}$$

Hence, an ideal element could theoretically be defined that would only represent the above effect of induced magnetic field and associated opposite voltage (such element is called an inductor). If such an element is connected to the already defined ac voltage source, the current flowing through the inductor would be the integral of a sine waveform, which is a cosine with a negative sign, or equivalently a sine that is shifted $-\pi/2$ radians. In power systems engineering it is usually said that the current waveform across an inductor lags behind the voltage waveform.

A magnitude called inductive reactance (X_L) can be defined based on the above, which represents the quotient of the amplitudes of the voltage sine waveform and the current shifted-sine waveform (obviously, as a ratio between the voltage and the current, it is measured in ohms). Inductive reactance then is a measure of the impediment to the flow of currents that are caused by induced electromagnetic fields. For a given inductor of value L, inductive reactance is directly proportional to frequency.

Similarly to the induced voltage concept, it can be shown that the time rate of change of magnetic fields induces electric fields.

$$v(t) = \frac{1}{C}\int i(t)\,dt$$

For an ideal element that would only represent the effect of induced electric field, which we would call a capacitor, that connects to the already defined ac voltage source, the current flowing through the capacitor would be the first derivative of a sine waveform, which is a cosine or equivalently a sine that is shifted $+\pi/2$ radians. In power systems engineering it is usually said that the current waveform across a capacitor leads the voltage waveform.

A magnitude called capacitive reactance (X_C) can be defined based on the above, which represents the quotient of the amplitudes of the voltage sine waveform and the current shifted-sine waveform (obviously, as a ratio between the voltage and the current, it is measured in ohms). For a given capacitor of value C, capacitive reactance is inversely proportional to frequency.

3.1.4.4 Load and Impedance

A general relation between **V** and **I** can be defined through a magnitude called impedance, which represents, in the complex plane, the obstacle/impediment that a certain load poses to the flow of current through it for a given voltage. Impedance has two components: a real component called resistance as defined in Section 3.1.4.2 and an imaginary component called reactance as defined in Section 3.1.4.3, which would add up all the effects of inductive and capacitive reactances. Impedance is also measured in ohms.

$$Z = \frac{V}{I}$$

Impedance is a fundamental concept for power systems engineering and also for telecommunications, as it provides the most generic way of representing the electromagnetic energy flowing through a certain system. In electrical terms, the concept *load* is normally used as a substitute of impedance, to define a combination of resistors, capacitors, and inductors as per the above definitions.

3.1.5 Power

Power represents the rate at which energy is flowing or work is being done. If current is the amount of charge that flows per unit of time, and voltage was defined as electric potential energy per unit of charge, multiplying both magnitudes cancels the charge and provides energy per unit of time, that is, power (or, more precisely, instantaneous power).

$$p(t) = v(t)i(t)$$

In the previous example of a dc system, the power that is being delivered to the resistor at any time is constant.

$$P = I^2 R$$

For ac systems the power depends heavily on the load (impedance). The general case can be studied considering the instantaneous power as the product of voltage and current. The instantaneous power delivered at any time t is a sinusoid waveform that is double the system frequency, shifted along the x axis.

$$p(t) = \sqrt{2}V \sin(2\pi f_S t) \cdot \sqrt{2}I \sin(2\pi f_S t - \varphi)$$
$$= VI \cos\varphi - VI \cos\left[2\pi(2f_S)t - \varphi\right]$$

The instantaneous power is analyzed under the following power concepts:

- Real power: Sometimes called active power or just power; the average value of the instantaneous power, equal to $VI \cos\varphi$. Real power is measured in watts (W).
- Reactive power: It represents back and forth movement of power among reactances, equal to $VI \sin\varphi$. Reactive power is measured in volt-amperes reactive (var).

- Apparent power: In an ac system the value of the apparent power is $V \cdot I$ (i.e., the product of rms voltage and rms current). Apparent power is always greater than or equal to real and reactive power and is measured in volt-amperes (VA).

Real power in an electric system represents the average value of instantaneous power and is power that actually does work. Reactive power is power in the system that is moving back and forth and does not do useful work. Apparent power value is used to thermally rate electrical devices normally present in power systems (conductors, transformers, and so forth).

The objective of a transmission or distribution system is to maximize the power transferred, minimizing the losses. The first source of losses is the resistive part of power line impedances. The lower the resistive part of the impedance, the lower the losses. For a purely resistive value, the average dissipated power is $I^2 R$. Hence, losses in a transmission line can be reduced by increasing the transmission voltage, which then allows the current to be reduced proportionally and the power dissipation to be reduced with the square of the current decrease. This explains why long transmission lines employ high voltage.

The other loss source is the reactive power. A term called cos φ, power factor of the load, is defined. The power factor should ideally be equal to 1, and it has a very important effect on the values of real and reactive powers. For two systems delivering the same amount of real (active) power, the system with the lower power factor will have higher circulating currents due to power that returns to the source and hence higher losses. This is why utilities generally try to maintain power factors close to 1 (voltage and current almost in phase) so that most of the power that is flowing is doing useful work. The way to achieve this when loads are typically inductive is through capacitor banks that are normally connected near the large inductive loads (e.g., long power lines) to compensate their reactive power. This is called line compensation.

3.1.6 Energy

Electromagnetism is defined on the basis of a fundamental property of matter such as electric charge that has the capacity of driving forces on other charged particles. Thus, the concepts of energy and work can be introduced.

Energy is the ability to perform work. Energy cannot be created or destroyed, but can be converted from one form to another [7]. Electrical energy is usually measured in watt-hours (Wh) so the relation between power and energy is immediately seen as the integral function of power with respect to time.

Electricity is a secondary energy source, sometimes referred to as an energy carrier: customers get electricity from the conversion of other (traditional) energy generation sources. These sources of energy are called primary energy sources. When reaching final customers, electricity usually undergoes another conversion that allows for it to be presented for useful purposes: charging electric batteries or powering electric motors that will ultimately be used to perform work in the form of motion, lighting, and so forth.

3.2   Structure of Electric Power Systems

From a general perspective [8], an electric power system is usually understood as a very large network that links power plants (large or small) to loads, by means of an electric grid that may span a whole continent, such as Europe or North America. A power system thus typically extends from a power plant right up to the sockets inside customers' premises. These are sometimes referred to as full power systems as they are autonomous.

Smaller power systems could be made of part or sections of a larger, full system. Figure 3.4 shows several elements that operate together and are connected to a power supplying network.

The subsystem represented in Figure 3.4(a) could be one of a final user of the electric energy of a full power system. The subsystem represented in Figure 3.4(b) could be one of a small power plant working as distributed generation (DG). Most of these power systems operate only when connected to a full power system. Power systems that are supplied by an external electricity source

(a) Electric motor connected to power grid

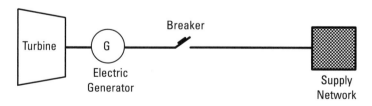

(b) Electric generator connected to power grid

Figure 3.4   (a, b) Specific purpose power subsystems.

or that produce (by conversion from other sources) electricity and convey it to a larger grid are called partial power systems.

The power systems that are of interest for our purposes are the large scale, full power systems that span large distances and have been deployed over decades by power companies. They consist of the segments introduced in Chapter 1. Generation is the production of electricity at power stations or generating units where a form of primary energy is converted into electricity. Transmission is the network that moves power from one part of a country or a region to another; it is usually a well-interconnected infrastructure in which multiple power lines link different substations, which change voltage levels, offering enhanced redundancy. Distribution finally delivers the power (we could say *locally* when compared to the transmission system) to the final loads (a majority of which are supplied at low voltage) via intermediate steps at which the voltage is converted down (transformed) to lower levels. The distribution system ends up at the energy consumption points or loads where power is used for its final purpose.

From a high-level perspective, it could be said that power systems in different parts of the world have converged over the past century towards a similar structure and configuration. However, within the same country differences usually exist among regions, among areas served by different companies or even areas served by the same company.

From a regulatory perspective, some power systems throughout the world are still operated as government monopolies and others as independent transmission system operators (TSOs) or distribution system operators (DSOs). There are parts of the world in which the deregulation and privatization of the industry has already completely changed the industry landscape, while in others the impact is still to be seen.

3.2.1 Power Generation

Power plants convert the energy stored in the fuel (mainly coal, oil, natural gas, enriched uranium) or renewable energies (water, wind, solar) into electric energy.

Conventional modern generators produce electricity at a frequency that is a multiple of the rotation speed of the machine. Voltage is usually no more than 6 to 40 kV. The power output is determined by the amount of steam driving the turbine, which depends mainly on the boiler. The voltage of that power is determined by the current in the rotating winding (i.e., the rotor) of the synchronous generator. The output is taken from the fixed winding (i.e., the stator). The voltage is stepped up by a transformer, normally to a much higher voltage. At that high voltage, the generator connects to the grid in a substation.

Traditional power plants generate ac power from synchronous generators that provide three-phase electric power, such that the voltage source is actually a combination of three ac voltage sources derived from the generator with their respective voltage phasors separated by phase angles of 120°. Wind turbines and mini hydro units normally employ asynchronous generators, in which the waveform of the generated voltage is not necessarily synchronized with the rotation of the generator.

DG refers to generation that connects into the distribution system, as opposed to conventional centralized power generation systems. The Electric Power Research Institute (EPRI) [9] has defined distributed generation as the "utilization of small (0 to 5 MW), modular power generation technologies dispersed throughout a utility's distribution system in order to reduce T&D loading or load growth and thereby defer the upgrade of T&D facilities, reduce system losses, improve power quality, and reliability." Small generators are constantly improving in terms of cost and efficiency, becoming closer to the performance of large power plants.

3.2.2 Transmission Systems

Power from generation plants is carried first through transmission systems, which consist of transmission lines that carry electric power at various voltage levels. A transmission system corresponds to a networked, meshed topology infrastructure, connecting generation and substations together into a grid that usually is defined at 100 kV or more. The electricity flows over high-voltage (HV) transmission lines to a series of substations where the voltage is stepped down by transformers to levels appropriate for distribution systems.

Preferred ac rms voltage levels are internationally standardized in IEC 60038:2009 as:

- 362 kV or 420 kV; 420 kV or 550 kV; 800 kV; 1,100 kV or 1,200 kV highest voltages for three-phase systems having a highest voltage for equipment exceeding 245 kV.

- 66 (alternatively, 69) kV; 110 (alternatively, 115) kV or 132 (alternatively, 138) kV; 220 (alternatively, 230) kV nominal voltages for three-phase systems having a nominal voltage above 35 kV and not exceeding 230 kV.

- 11 (alternatively, 10) kV; 22 (alternatively, 20) kV; 33 (alternatively, 30) kV or 35 kV nominal voltages for three-phase systems having a nominal voltage above 1 kV and not exceeding 35 kV. There is a separate set of values specific for North American practice.

- In the case of systems having a nominal voltage between 100V and 1,000V inclusive, 230/400V is standard for three-phase, four-wire systems (50 Hz or 60 Hz) and also 120/208V for 60 Hz. For three-wire systems, 230V between phases is standard for 50 Hz and 240V for 60 Hz. For single-phase three-wire systems at 60 Hz, 120/240V is standard.

Although not explicitly mentioned in IEC 60038:2009, low voltage (LV) was defined as "a set of voltage levels used for the distribution of electricity and whose upper limit is generally accepted to be 1,000 V for alternating current" (IEC 60050-601:1985/AMD1:1998, 601-01-26), while HV is either "the set of voltage levels in excess of low voltage" or "the set of upper voltage levels used in power systems for bulk transmission of electricity" (IEC 60050-601:1985/AMD1:1998, 601-01-27). Medium voltage (MV) as a concept is not used in some countries (e.g., United Kingdom and Australia), it is "any set of voltage levels lying between low and high voltage" (IEC 60050-601:1985/ AMD1:1998, IEV 601-01-28) and the problem to define it is that the actual boundary between MV and HV levels depends on local practices. Setting it at 35 kV is a good compromise based on IEC 60038:2009 standard voltages.

In Europe, overhead transmission lines are used in open areas such as interconnections between cities or along wide roads within the city. In congested areas within cities, underground cables are used for electric energy transmission. The underground transmission system is environmentally preferable but has a significantly higher cost.

Transmission lines are deployed with three wires along with a ground wire. Virtually all ac transmission systems are three-phase transmission systems.

3.2.3 Distribution Systems

Distribution segment is widely recognized as the most challenging part of the smart grid due to its ubiquity. Voltage levels of 132 (110 in some places) or 66 kV are usual HV levels that can be found in (European) distribution networks. Voltages below that (e.g., 30, 20, 10 kV) are commonly found in MV distribution networks. Distribution levels below 1 kV are within what is known as LV.

MV grid topologies can be classified in three groups:

- Radial topology: Radial lines are used to connect primary substations (PSs) with secondary substations (SSs), and the SSs among them. These MV lines or "feeders" can be used exclusively for one SS or can be used to reach several of them. Radial systems keep central control of all the SSs. These radial topologies show a tree-shaped configuration when they grow in complexity; they are a less expensive topology to develop, operate, and maintain, but they are also less reliable.

- Ring topology: This is a fault-tolerant topology to overcome the weakness of radial topology when there is a disconnection of one element of the MV line that interrupts electricity service (outage) in the rest of the connected substations. A ring topology is an improved evolution of the radial topology, connecting substations to other MV lines to create redundancy. Independently of the physical configuration, the grid is operated radially, but on the event of a failure in a feeder, other elements are maneuvered to reconfigure the grid in such a way that outages are avoided.

• Networked topology: Networked topology consists of primary and secondary substations connected through multiple MV lines to provide a variety of distribution alternatives. Thus, the reconfiguration options to overcome faults are multiple, and in the event of failure, alternative solutions may be found to reroute electricity.

LV distribution systems can be single-phase or three-phase. In Europe, for example, they are usually three-phase, 230V/400V systems (i.e., each phase has an rms voltage of 230V and the rms voltage between two phases is 400V).

LV grids present more complex and heterogeneous topologies than MV grids. The exact topology of LV systems depends on the extension and specific features of the service area, the type, number and density of points of supply (loads), country- and utility-specific operating procedures, and range of options in international standards. An SS typically supplies electricity to one or several LV lines, with one or multiple MV-to-LV transformers at the same site. LV topology is typically radial, having multiple branches that connect to extended feeders, but there are also cases of networked grids and even ring or dual-fed configurations in LV networks. LV lines are typically shorter than MV lines, and their characteristics are different depending on the service area.

3.2.4 Distribution System Elements

3.2.4.1 Substations

Primary Substations

A PS (Figure 3.5) transforms voltage from HV to MV by means of transformers. It basically consists of a complex set of circuit breaking, voltage transforming, and control equipment arranged to direct the flow of electrical power.

Transformation may take place in several stages and at several substations in sequence, progressively reducing to the voltage required for transformation at SSs.

A substation that has a step-down transformer will decrease the voltage while increasing the current for residential and commercial distribution.

The functions of a PS can be gathered into three categories:

Figure 3.5   Primary substation.

- Safety: It separates those parts where an electrical fault can take place from the rest of the system.
- Operation: It minimizes energy losses and also allows separating parts of the network to perform the necessary maintenance functions on the equipment, or to install new equipment.
- Interconnection: It allows interconnection of different electrical networks with diverse voltages, or interconnection of several lines with the same voltage level.

PSs are designed for several specific functions such as regulating voltage to compensate for system changes, switching transmission and distribution circuits, providing lightning protection, measuring power quality and other parameters, hosting communication, protection, and control devices, controlling reactive power, and providing automatic disconnection of circuits experiencing faults.

A PS will contain line termination structures, switchgear, transformer(s), elements for surge protection and grounding, and electronic systems for protection, control, and metering. Other devices such as capacitors and voltage regulators are also usually located in these premises. Pieces of equipment are usually connected to each other through conductor buses or cables.

Secondary Substations

SSs are those premises at the end of MV networks where electricity is transformed to LV. Stemming from the SSs, LV grids are deployed to reach the customers; SSs are located close to the end-users. In Europe, SSs normally supply at LV an area corresponding to a radius of some hundred meters from the SS. North and Central American systems of distribution consist of an MV network from which numerous (small) MV/LV distribution transformers each supply one or several consumers, by direct service cable (or line) from the SS transformer location.

SS transformers reduce voltage to levels that are then distributed to residential, commercial, and small and medium industrial customers. Figure 3.6 shows a general electrical scheme of an SS.

The main parts of an SS are:

- MV lines: These are the power lines supporting specific voltages and currents that originate from primary substations and deliver electricity to SSs.

- Switchgear or MV panels: They are the interface between MV lines and transformer(s). They serve two main purposes: protection (to connect, disconnect, and protect the transformer) and line interconnection (to guarantee continuity and allow for operation of MV feeders). Old switchgears were air insulated. Modern switchgears use some sort of gaseous dielectric medium (most common today is sulfur hexafluoride, SF6).

- Transformer: The device that reduces the voltage from MV levels to LV.

- LV panel: It is the element located beyond the transformer, connected to its secondary winding, which divides the total power in a number of LV feeders. It is usually a large single panel, frame, or assembly of panels, composed of four horizontal bars where the feeders are connected (for the three phases and the neutral). The LV switchboard can have switches, overcurrent, and other protective devices.

- LV feeders: These are the power lines supporting specific voltages and currents that deliver electric energy to customers. Usually several LV lines come out of a single SS, which then provide electric service to buildings and premises around.

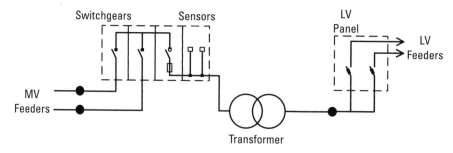

Figure 3.6   Secondary substation electrical scheme.

SSs may be located at different places depending on the electrical topology of the grid and the needs in each case. They can be classified into two main groups according to the location of the MV/LV transformer:

- Indoor, either in shelters or underground;
- Outdoor, usually with overhead transformers on poles or a similar configuration.

Indoor Secondary Substations   The most common indoor locations where SSs may be found are (Figure 3.7):

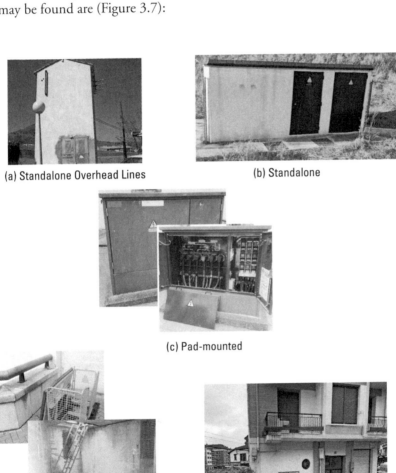

(a) Standalone Overhead Lines   (b) Standalone

(c) Pad-mounted

(d) Underground   (e) Building

Figure 3.7   Common types of indoor secondary substations.

- Standalone shelters: In this case, MV lines may reach the SS either overhead or underground. In both cases, internal structures are very similar (the only differences are at the MV lines entry point and its supporting elements). A special case would be the compact SSs, or the pad-mounted transformers. These are usually large cabinets installed in publicly accessible areas, which comprise all different operational modules (MV, LV, and transformer) within a single isolated and compact structure.

- Underground: These SSs are fed by underground MV cables.

- Integrated as part of larger commercial or residential buildings: Secondary substations belong to the building itself and are usually located on the lowest floors inside a building, at ground level or at the basement. The structure and operation are the same as previous cases.

In general, indoor SSs comprise the following elements:

- Enclosure: The protected place where all the SS elements are located. It may either be a building, a prefabricated shelter, and so forth.

- Transformer(s), switchgear, and LV switchboard: Described in Section 3.2.5.

- Grounding: It is common to find two grounds, one for MV and the other one for LV. The MV grounding connects to MV cables shield and metallic parts of every element in the transformer tank, switchgear, and LV switchboard enclosures. LV grounding connects to the LV neutral. Hazardous situations may arise from this ground separation.

Outdoor Secondary Substations   This type of SS is typical in rural and suburban areas in Europe and especially in the United States. The outstanding representative element of these SSs is the transformer itself. Common outdoors locations are (Figure 3.8) on a structure or on a pole.

3.2.4.2   Distribution Power Lines

Medium Voltage Lines

MV lines are the elements that carry electric power in the MV distribution grid. Cables must be selected to safely provide adequate electric power with continuous operation, in a distribution system that can withstand overload conditions and/or unexpected demand.

There is a wide variety of cables and many classifications may be done. For instance, they can be classified according to the material of the conductor part. The most widely used metals are aluminum and copper (copper is highly conductive; aluminum is lighter and cheaper). Another classification may

Figure 3.8   Common types of outdoor secondary substations.

consider the section of the cable. Depending on the power carried, the section of the cables varies (higher power requires larger sections). The bundling is also important; there are two possibilities: single core and three core cables. Three core cables have three conductors in the same cable, one conductor for each phase inside. Single core cables have one phase conductor for each cable; therefore, three cables are necessary for each MV line. Single core cables are preferred for long stretches. Finally, the most common and straightforward classification is referred to the location where the cables are laid: overhead or underground.

Overhead Lines   Overhead lines are laid on poles. These poles can be made of wood, steel, concrete, or fiberglass. Different configurations can be found: one circuit per pole or more than one circuit.

Overhead lines are normally classified according to the conductor. Today, utilities use aluminum for almost all new overhead installations; however, many copper overhead lines are still in service.

There are two major categories of overhead conductors: homogeneous and nonhomogeneous. The first category includes copper, all-aluminum conductor (AAC), and all-aluminum alloy conductor (AAAC). The second category includes more types; the usual ones are aluminum conductor steel reinforced (ACSR), aluminum conductor alloy reinforced (ACAR), aluminum conductor steel supported (ACSS), or aluminum alloy conductor steel reinforced (AACSR). The use of steel greatly increases the mechanical properties (strength) of the cable.

The wire can be insulated or not insulated. The bare-wired (uninsulated) is the most common type of line used in overhead power lines. Covered wires are AAC, AAAC, or ACSR conductors covered with polyethylene (PE) or cross-linked polyethylene (XLPE). Because the lines are insulated, there is little chance of damage from vegetation or wildlife, so they exhibit improved reliability.

Underground Lines   Underground lines (Figure 3.9), hidden from view, are considered to be safer as there are fewer opportunities for accidental contact, statistically more reliable and with a lower maintenance cost. However, the installation cost is significantly higher than in overhead lines.

In general, the structure of an underground cable consists of the following parts (from inner to outer layers): conductor, conductor shield, insulation, insulation shield, neutral, or shield and jacket.

- Conductor: It is the element that carries electrical current, made of aluminum or copper. The conductor can be solid or stranded. Solid ones show better conductivity and are less expensive. However, stranded conductors present improved mechanical flexibility and durability.

- Conductor shield: It surrounds and covers the conductor. It provides for a smooth, radial electric field within the insulation. Without this shield, the electric field gradient would concentrate at the interface between the conductor and the insulation; the increased localized stress could adversely affect the insulation.

- Insulation: It is made of nonconductive material, and allows cables with small diameters to support conductors at significant voltage. According to the insulation, the cables can be PILC, or synthetic: different materials may be used as insulator such as polyvinyl chloride (PVC), PE, XLPE, or ethylene propylene rubber (EPR).

- Insulation shield: It surrounds and covers the insulation. As with the conductor shield, the insulation shield evens out the electric field at the interface between the insulation and the neutral/shield. The insulation

(a) XLPE Underground   (b) Three Core Cable

Figure 3.9   Underground cables.

shield is easily strippable in order to be terminated and spliced. Its materials are similar to those of the conductor shield.

- Neutral or shield (sheath): The shield is the metallic barrier that surrounds and covers the insulation shield. The neutral or shield holds the outside of the cable at ground potential and provides a path for return current and also for fault current. The shield protects the cable from lightning strikes and from current from other fault sources. It can be built with wires or tapes.

- Jacket: It is the outer cover of the cable. Its purpose is to protect all inner elements from mechanical damage, chemicals, moisture, and exposure to harmful environmental conditions. Most jackets are made of extrudable plastics, and PVC and PE are common materials today.

Underground cables can be single-core or three-core (each core with its own insulation and shield, but sharing the outer jacket). Both types can have armored layers over the insulation to provide the cable with additional mechanical protection.

Low-Voltage Lines

LV lines are the elements that carry electric power in the LV distribution grid, bringing electric energy from SSs to individual LV customers. In European countries, the LV grid is usually a larger infrastructure than the MV grid, and it requires high investment to map, document, and correctly manage the entire LV infrastructure. Thus, it is not unusual to find gaps in utilities' records with respect to the exact route specific LV lines follow, which customers connect to each of the LV lines that depend on an SS, or to which phase every LV customer is connected.

Usually in European SSs, the output from a transformer is connected to LV panels via a switch, or simply through isolating links. LV panels are usually four- to 12-way, three-phase, four-wire distribution fuse boards or circuit-breaker boards. They control and protect outgoing four-core LV lines. In densely loaded areas, a standard length of LV line is laid to form a LV grid (e.g., with one cable along each pavement and two-, three-, or four-way link boxes located in manholes at street corners, where two or more cables cross). Links are inserted in such a way that LV lines form radial circuits from the SS with open-ended branches.

Link boxes are installed underground, but it is also usual to find free-standing LV distribution pillars, placed above ground at certain points in the grid, or LV distribution cabinets. These weatherproof cabinets are installed either against a wall or, where possible, built into the wall (see Figure 3.10).

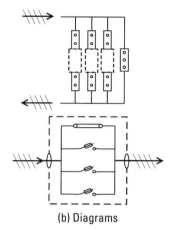

(a) Fuse Box   (b) Diagrams

Figure 3.10   Built-in LV cabinets.

LV distribution cabinets located at the entrance of buildings and houses are also the elements that usually delimit the edge of utility grid, such that LV infrastructure beyond the cabinet is the ownership and responsibility of customers. Part of this infrastructure is usually the place where electricity meters are located. Modern buildings usually have a specific sheltered room to concentrate the meters of all customers in a single place. These rooms host the general breaker that can isolate the customer installation from the LV grid, a general busbar with fuses that protect the individual customer circuits, all the meters, and the individual supply lines that go to customers.

North and Central American practice differs fundamentally from that in Europe, as utilities' LV grids are practically nonexistent. The distribution is effectively carried out at MV, such that the MV grid is, in fact, a three-phase, four-wire system from which single-phase distribution networks (phase and neutral conductors) supply numerous single-phase transformers, the secondary windings of which are center-tapped to produce LV single-phase, three-wire supplies. Each SS (MV/LV distribution transformer) normally supplies one or several premises directly from the transformer position by radial service cable(s) or by overhead line(s). Many other systems exist in the Americas, but the one described is the most common.

Similarly to MV lines, the LV lines are basically divided into overhead and underground. However, materials and dimensions differ greatly for various types of cables, even more than in MV.

LV overhead lines use either bare conductors supported on glass/ceramic insulators or an aerial bundled cable system. Both bare conductors and bundled cables can actually be laid outdoor on poles or wall-mounted. Conducting material is usually either aluminum or copper (the former is preferred).

Medium to large-sized towns and cities have underground cable distribution systems. These also show a typical structure of conductor that is insulated with the same materials discussed for MV and is protected by an outer PVC jacket. Underground cables can be found inside utility tunnels, laid in ducts or tubes, or directly buried in trenches.

Table 3.1 summarizes the main features of European LV grids.

3.2.5   Components of System Elements

3.2.5.1   Transformers

Transformers reduce the voltage of an electric utility power line to lower voltages or vice versa. They are used in different parts of the electric power system, such as generation, transmission, and distribution. This section focuses on MV/LV transformers due to their interest for the smart grid.

MV/LV transformers or distribution transformers reduce the voltage of an electric utility power MV distribution line to lower voltages suitable for customer equipment. The transformer primary is the winding that draws power, and the transformer secondary winding delivers power.

Basically, a transformer consists of two primary components: a core made of magnetically permeable material and conductors made of a low resistance material such as copper or aluminum. The conductors are wound around the core at differing ratios, transforming current from one voltage to another. This process requires a liquid insulation material (or air for smaller transformers) to cool and insulate.

In general, the MV/LV transformers can be classified according to cooling technology or connection type. Cooling technology refers to the technology that is used to cool the transformer. According to the coolant, the transformers can be classified into liquid-immersed or dry-type.

Table 3.1
Main Features of European LV Networks

Feature	High-Density Residential Area	Low-Density Residential Area
Type of SS	Underground or integrated in building	Standalone shelter or pole
Transformers per SS	1–4	1
Average number of LV customers per SS	150–300	5–100
LV feeders per SS	4–20	1–6
Average length of LV lines	100m	300m
Type of LV line	Underground	Overhead

Source: [10].

In terms of connection type, transformers are usually connected to one or three MV phases (see Figure 3.11; two-phase transformers also exist).

3.2.5.2   MV Switchgear

Usually the term switchgear (Figure 3.12) is referred to a complex device with the purpose of protecting/isolating and interconnecting different elements such as feeders and transformers. It is a general term covering switching and interrupting devices and their combination with associated control, metering, protective, and regulating devices and also assemblies of these devices with associated interconnections, accessories, enclosures, and supporting structures.

Switchgears can be classified according to their design and the insulation material. According to the design there are different types of switchgears:

- Masonry switchgear: This type of switchgear is a space enclosed by masonry walls on three sides and by grounded removable metal barriers on its front side. The barriers allow internal access for maintenance tasks. Inside the switchgear there are MV cables, their corresponding bushings, fuses, breakers, and supporting elements. The insulator element is simply the air. This type is now considered obsolete and is being replaced gradually by modern elements ones:

- Prefabricated switchgear: This switchgear is provided already built from factory. The interconnection between equipment and the cabling is done while it is installed in the secondary substation. A further classification is found in IEEE Standard C37.100-1992:

Figure 3.11   Three-phase transformer.

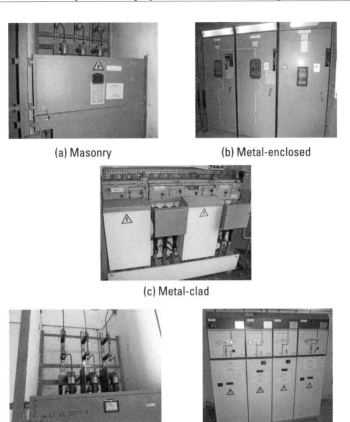

(a) Masonry   (b) Metal-enclosed

(c) Metal-clad

(d) Air   (e) SF6

Figure 3.12   MV switchgear types.

• *Metal-enclosed switchgear:* All the equipment are placed inside a metal case or housing, normally grounded.

• *Metal-clad switchgear:* This is a specific type of metal-enclosed switchgear, in which major parts of the primary circuit are completely enclosed by grounded metal barriers that have no openings between compartments, all live parts are enclosed within grounded metal compartments, and primary bus conductors and connections are covered with insulating material throughout.

In terms of insulating material, there are different types of switchgears:

• Air-insulated switchgear: The air is the insulating substance. They can be prefabricated switchgear or masonry.

- Gas (SF_6) insulated switchgear: This is an evolution from the prefabricated air-insulated switchgear. Air is replaced by sulfur hexafluoride gas. It is a nontoxic, inert, insulating, and cooling gas of high dielectric strength and thermal stability. This results in considerable space and weight savings and improvements in the operational safety.

The main components of switchgears are switches, breakers, and bushings.

3.2.5.3   Circuit Breakers and Switches

Circuit breakers and switches are similar devices. A circuit-breaker is (IEC 60050-441:1984/AMD1:2000, 441-14-20) a "mechanical switching device, capable of making, carrying and breaking currents under normal circuit conditions and also making, carrying for a specified duration and breaking currents under specified abnormal circuit conditions such as those of short circuit." The circuit breaker consists of power contacts with arc clearing capability and associated control and auxiliary circuits for closing and tripping the breaker under the required conditions.

The switch (IEC 60050-441:1984/AMD1:2000, 441-14-10) provides similar functions to the circuit breaker, but cannot break short-circuit currents.

Both circuit breakers and switches have a similar physical principle: they interrupt current during a zero-crossing of the signal (i.e., a 50/60-Hz sinusoidal waveform at a certain voltage level). When a switch is opened, an arc is created. Each half-cycle, the ac current momentarily stops as the current is reversing directions. During this period, when the current is changing, the arc is not conducting and starts to de-ionize. If the dielectric strength builds up faster than the recovery voltage, the arc finishes and finally the circuit is opened.

Depending on the operation, switching can be divided into manual or automatic (by adding a motor actuator and controller). Depending on the insulator used, there are air pressured, vacuum, oil, and SF_6 models. SF_6-based switches and breakers are replacing all others, which have been used in the industry for decades.

3.2.5.4   Bushings/Insulators

A bushing is (IEC 60050-471:2007/AMD1:2015, 471-02-01) a "device that enables one or several conductors to pass through a partition such as a wall or a tank, and insulate the conductors from it." Bushings may be of many different types (e.g., liquid, gas, impregnated paper, ceramic, glass, resin). The general purpose of a bushing is to transmit electrical power in or out of enclosures of an electrical apparatus such as transformers, circuit breakers, shunt reactors, and power capacitors.

3.2.5.5 Fuses

Fuses (see Figure 3.13) are one of the oldest and most simple protective devices. A fuse is (IEC 60050-441:1984/AMD1:2000, 441-18-01) "a device that by the fusing of one or more of its specially designed and proportioned components opens the circuit in which it is inserted by breaking the current when this exceeds a given value for a sufficient time."

When the current flowing through the fuse exceeds a predetermined value, the heat produced by the current in the fusible link melts it and interrupts the current. Because the current must last long enough for the link to melt, fuses have inherently a time delay. Fuses are relatively economical, and they do not need any auxiliary devices such as instrument transformers and relays. Their one disadvantage is that they are destroyed in the process of opening the circuit, and then they must be replaced. MV fuses can be found in many locations, along with breakers, transformers, MV lines, and so forth.

3.3 Operations of Power Systems

Power system operations need to consider that production and consumption has to be always in balance. Whenever there is a mismatch between supply and demand in a large system, the overall dynamic balance may become compromised. The operation of the power system ensures that electricity is delivered to consumers while keeping the system stable and protected. System instability refers to voltage and frequency values outside predefined thresholds.

The management of a power system is a combination of planning, delivery, operation, and maintenance. These processes have both short-term and long-term components. As the system consists of the generation, transmission, distribution, and supply segments, those processes will exist across all of them.

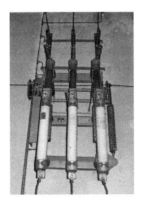

Figure 3.13 Fuses.

There are two functions that can be considered transversal to all grid management areas: protection and control. They apply to hierarchically organized systems that have different vision levels of the whole power system. Operations depend on a combination of automated or semiautomated control, and actions requiring direct human (system operator) intervention. Operations are assisted by all electromechanical elements developed and installed during the life of the electric grid, and recently enhanced with the support of information and communication technologies (ICTs):

- Protection: This function intends to ensure the general safety of the system and its elements. Protection schemes must act immediately (real time) when there is a condition that might cause personal injuries to general public, or equipment damage. The faulty condition is detected and located based on voltage and current measurements, but also on some other parameters. A fundamental part of the operation of a power system is to quickly detect and clear faults. Protection cannot avoid disturbances in the system, but can minimize the consequences through rapid, secure, and selective disconnection of the faulty equipment, and automatic reclosing for supply recovery in case of transient failures.

- Control: Power system operators manage their grids from utility control centers (UCCs). Many types of different UCCs exist, and they may deal with different grid domains. Most of the routine operations of a well-designed system should not require any human intervention. However, a certain number of manual operations are needed. In normal operating conditions, the collection of operating data is used to analyze system performance for planning and contingency analysis. This is usually done automatically.

The objective of producing the power needed in the system is a task for the generation segment. In the traditional monopolistic environment before the deregulation of electricity markets, vertically integrated utilities controlled all the power system domains and knew when, where, and how much electricity was going to be needed. Scheduling of energy production was relatively easy. Even if different generation sources had different constraints (e.g., costs, speed of startup and shutdown, capacity factor), all data was available for the utility. The unit commitment [4] assigns a production rate and temporal slot sometime in advance of the real need; the reserves (i.e., the generation units that must remain available in case of contingencies) need to be planned too, and they have costs.

After market deregulation, operation became more challenging because generation turns into an open market where energy producers are usually independent entities that offer their production capacity and get it awarded. There is usually a wholesale bidding process and a price assigned per MW to the different generation sources.

Frequency stability is a fundamental concept for power system operation that concerns generation as well. Traditional production of electricity (hydro, fuel, nuclear) involves mechanical elements (e.g., as with steam, water, or gas flows through a turbine). Thus, mechanical power has an effect on the rotational speed of the turbine and consequently in the exact frequency of the electricity signal. If rotational speed is higher, frequency is higher too. However, effects on frequency also come from the load connected to the system: when load is heavy, the turbine will tend to rotate more slowly and the output frequency will be lower. This effect needs to be compensated on the generator side with more fuel to keep the frequency as close as possible to the 50- or 60-Hz nominal value.

Voltage levels need to be controlled too. Loads in the system exhibit a reactive behavior, so if the total consumption of reactive power is high, the generator output power will not be efficiently used. Keeping reactive consumption low maximizes the real power flowing in the system. The grid (transmission and distribution) takes care of compensating loads; otherwise, generators need to act.

From an operational perspective, the transmission grid must ensure power lines support the electricity transported. This means dynamically adapting to physical limitations and tolerances (e.g., thermal), with the aim of maximizing availability and minimizing system losses. Control of the reactive part of the load is done with volt-ampere reactive (var) regulation elements (inductors, capacitors, and semiconductor switches) deployed in the grid. Market deregulation adds further challenges to transmission grid operations since network capacity expansion must be coordinated with new actors that build generation plants to be connected to the grid.

Distribution systems have a fundamental duty to control power quality (e.g., voltage and frequency parameters) close to the end customer. Key aspects of distribution operations are voltage regulation (e.g., to keep LV levels between certain thresholds both in the unloaded and full-load conditions), power factor (i.e., to control the reactive part of the load), frequency, and harmonics (due to the growing presence of solid-state switching devices in the grid) and voltage unbalance among phases in multiphase systems.

Maintenance of power systems is not conceptually different to the practices of other businesses. There is a need for reactive maintenance when an unplanned failure occurs and also for preventive (proactive) maintenance to minimize future system failures. Use of ICTs is relevant to collect information and perform operation and maintenance (O&M) more effectively by analyzing thousands of complex parameters that may give early indication of problems due to degradation.

System planning is also part of operations. It tries to understand and forecast load needs and their location to adapt the grid consequently. System planning affects all system components and usually considers a short and a long-term scope. Long-term requires a system model and considers load evolution and associated changes needed in the existing system (e.g., new power lines or substations, refurbishment of existing infrastructure: feeders, transmission capacity) along with relevant constraints (e.g., economic) to develop different scenarios and adapt to them. Short-term planning implies detailed analysis of the infrastructure. In the distribution segment this is translated into different studies to analyze the grid itself: voltage drops (to identify weak points in the grid), sectionalizing study (to minimize outages), conductor analysis (to determine if existing ones are adequate or not), and power factor correction (e.g., by installing compensating elements in the grid) among others.

References

[1] National Academy of Engineering (NAE), "About NAE," 2016. Accessed March 21, 2016. https://www.nae.edu/About.aspx.

[2] Constable, G., and B. Somerville, *A Century of Innovation: Twenty Engineering Achievements That Transformed Our Lives*, Washington, D.C.: Joseph Henry Press, 2003.

[3] Oerter, R., *The Theory of Almost Everything: The Standard Model, the Unsung Triumph of Modern Physics*, New York: Plume, 2006.

[4] Kassakian, J. G., and R. Schmalensee, *The Future of the Electric Grid*, Cambridge, MA: Massachusetts Institute of Technology (MIT), 2011. Accessed March 25, 2016. http://mitei.mit.edu/publications/reports-studies/future-electric-grid.

[5] Budka, K. C., J. G. Deshpande, and M. Thottan, *Communication Networks for Smart Grids. Making Smart Grid Real*, London: Springer-Verlag, 2014.

[6] Liu. Z., *Ultra-High Voltage AC/DC Grids*, Waltham, MA: Academic Press, 2015.

[7] U.S. Energy Information Administration (EIA), "Electricity Explained," 2016. Accessed March 21, 2016. http://www.eia.gov/energyexplained/index.cfm?page=electricity_home.

[8] Ceraolo, M., and D. Poli, *Fundamentals of Electric Power Engineering*, New York: Wiley-IEEE Press, 2014.

[9] *Integration of Distributed Resources in Electric Utility Systems: Current Interconnection Practice and Unified Approach*, Palo Alto, CA: Electric Power Research Institute (EPRI), 1998. TR-111489.

[10] EDF, Iberdrola, DS2, Main.net, *Deliverable D44: Report on presenting the architecture of PLC system, the electricity network topologies, the operating modes and the equipment over which PLC access system will be installed*, Open PLC European Research Alliance (OPERA), 2005. IST Integrated Project No 507667. Funded by the European Commission.

4

Smart Grid Applications and Services

Systems operators use applications to manage the power system. These applications use telecommunication services to access all the system end-points where any measurement, monitoring, or control capability is needed.

The concept of service, as it is used in this book, is very specific to telecommunications. Services are provided by telecommunication networks and/or systems, and end-user applications are built on top of these services. Regardless of the complexity of services, utilities build their applications on them. Applications include programs (with coded algorithms), databases, graphical user interfaces, and geographical information systems. All these elements reside in servers accessed by the users of the systems through client machines, within information systems.

4.1 Utility Information Systems

Power grid information systems can be classified according to the part of the grid that they support through specific applications. Applications are more often classified according to the traditional operational needs of utilities [1] (Figure 4.1):

- Operations management systems: They are responsible for the management of systems directly controlling the utilities' generation, transmission and distribution. They also manage infrastructure field assets. The following systems can be specifically mentioned:
  - Energy management systems (EMSs) are in charge of the supervision, analysis, optimization, simulation, and control of the utilities' transmission and generation assets.

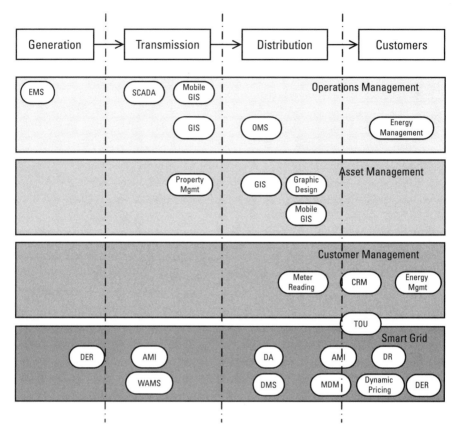

Figure 4.1 Overview of some utility applications [1].

- Distribution management systems (DMSs) are in charge of the monitoring, control, and analysis of the distribution network, providing support to operation awareness to improve safety and asset protection.
- Supervisory Control and Data Acquisition (SCADA) systems are real-time systems associated both to EMSs and DMSs to monitor and operate remote terminal units (RTUs) and intelligent electronic devices (IEDs).
- Outage management systems (OMSs) are used to manage the grid to help to restore power during service interruptions and to reduce the economic impact of power outages. OMSs analyze the location and extent of outages, can predict them through information from the distribution grid, and help dispatchers and crews to solve outages.

- Meter data management systems (MDMSs): These systems manage meter information both for grid and customer management.

- Customer management systems: These systems manage the relationship and the contact of utilities with their customers. Examples of these systems are customer relationship management (CRM) and energy management programs.

- Asset management systems: These manage work activities and utility data. Examples of these systems are those for work management, mapping, and geographic information system (GIS).

New applications associated to the smart grid must be preferably embedded in these systems. However, they can also have an entity as stand-alone new smart systems or applications.

4.2 Traditional Applications and Services

Utilities have historically made use of telecommunications either as a support for grid assets' operation or as the enabler for voice communications among operating staff. The basic objective of utility telecommunications is to connect remote premises and staff with central premises and among them.

4.2.1 Remote Control: SCADA Systems

SCADA refers to information and communication technology (ICT) systems deployed to manage (i.e., monitor and control) industrial systems, specifically in the context of the electric grid, remote electric power grid assets (substations in the case of transmission and distribution, power plants in the case of generation) from the utility control centers (UCCs). The basic element that provides control of the substations is the RTU. RTUs, programmable logic controllers (PLCs; not to be confused with power line communications), and IEDs are evolutions of the same element. According to [2], PLCs are different from RTUs in that they are not dumb telemetry systems, but may execute some control; PLCs are considered an intermediate step in the evolution of RTUs towards IEDs.

Neither SCADA systems nor IEDs are specific to electricity utilities as they can be found in different industrial environments. Of all the functions of a SCADA system, the one associated with control is the most critical one since it enables the operation of distant assets as if the operator was physically present. Data acquisition is needed for the purpose of taking informed decisions, and being able to check the status after a remote operation.

The elements inside a substation that need to be controlled are typically the transformer, the circuit breaker, and the switches. Voltage and current transformers are also monitored as inputs for decisions to take and outputs to be checked.

Some of the first SCADA systems were deployed [3] for generation and progressively spread to transmission first (EMS), and eventually distribution (DMS). First, SCADAs were all analog and now they are digital. When microcontrollers became available they enhanced and diversified the functions of devices deployed at substations. This allowed for the substitution of expensive wiring interfaces at substations with RTU interfaces through local substation telecommunication ports. At the same time, some utilities began deploying master stations into substations, so that they could operate independently for some automation functions and as slave devices for other functions, with the ultimate control being assigned to the UCCs. This diversity of devices and the protocols governing them were soon a source of confusion and cost. The IEEE Power and Energy Society Substations Committee coined a single term to identify them: IEDs as "any device incorporating one or more processors, with the capability to receive or send, data/control from, or to, an external source, for example electronic multi-function meters, digital relays, controllers" (IEC TS 61850-2:2003). Any device with a microprocessor and telecommunication capabilities in the context of power grids can be considered an IED.

SCADA system protocols are based on the IEDs being driven to accept commands (control points operation, analog output level setting, and request response) and to provide data from the field (status, analog, and accumulator data) to the SCADA master station. If the master station uses a serial protocol, it sequentially polls the IEDs. If the telecommunication system and protocols allow data to be sent spontaneously from the IEDs (e.g., IP-based systems), the polling is not needed. From the telecommunications perspective, SCADA systems based on polling strategies typically require a data delivery time below 200 ms (IEEE Standard Communication Delivery Time Performance Requirements for Electric Power Substation Automation, IEEE Standard 1646, 2004), which is demanding for certain telecommunication technologies.

Application-level protocols are relevant in the way that SCADA systems behave, perform, and evolve. The variety of protocols that have been used for SCADA is overwhelming. One RTU vendor reported having implemented 100 different protocols on its devices [3]. IEEE 1379-2000 (now withdrawn) covered a recommended practice for IEDs and RTUs in electric substations and consolidated two SCADA protocols above the rest: DNP3 and IEC 60870-5-101:2003 (both eventually evolved to be Ethernet/IP capable):

- The Distributed Network Protocol (DNP3) is a three-layer protocol (layers 1, 2, and 7 of the OSI model) specifically designed for data acquisition and control applications in the area of electric utilities [IEEE Standard for Electric Power Systems Communications-Distributed Network Protocol (DNP3), IEEE Standard 1815, 2012]. The DNP3 protocol is built on the framework specified by the IEC 60870-5 documents as a protocol subject to be implemented with commonly available hardware [4].

- IEC 60870-5 series is a set of protocols published by IEC TC 57 WG 3 "Telecontrol Protocols," chartered to develop protocol standards for telecontrol domain, teleprotection, and associated telecommunications for electric utility systems [other standards of the series are IEC 60870-5-102 (metering) and IEC 60870-5-103 (protection)] [5]. The two most popular standards are IEC 60870-5-101 and IEC 60870-5-104, based on the original IEC 60870-5-1 to IEC 60870-5-5. Additionally, the protocols include a user layer placed between the OSI application layer and the user's application program to add interoperability for functions such as clock synchronization and file transfers (Figure 4.2).

Nowadays, the evolution of SCADA systems is being mainly standardized in the IEC 61850 series, managed from IEC TC 57 WG 10 "Power System IED Communication and Associated Data Models." The IEC 61850 series was titled "communication networks and systems in substations" until 2005, indicating its scope was exclusively within substation IEDs. Starting in 2007, the scope of the series expanded outside the substation so the new title became "Communication Networks and Systems for Power Utility Automation" (including teleprotection and distributed generation).

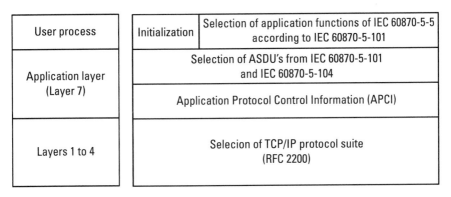

User process	Initialization	Selection of application functions of IEC 60870-5-5 according to IEC 60870-5-101
Application layer (Layer 7)	Selection of ASDU's from IEC 60870-5-101 and IEC 60870-5-104	
	Application Protocol Control Information (APCI)	
Layers 1 to 4	Selecion of TCP/IP protocol suite (RFC 2200)	

NOTE: Layers 5 and 6 are not used

Figure 4.2 Protocol stack of IEC 60870-5-104.

4.2.2 Teleprotection

The basic objective of teleprotection is to assist the protection system through the transfer of information to the locations where some element operation is possible, thus keeping the electricity infrastructure safe (e.g., substations on both ends of a power line) when there is an electrical fault. A fault is a flow of current that is large enough to cause damage to infrastructure or people. A fault is usually a short circuit that can happen between phase and ground or between two phases of a power line. It can be caused by the contact of one phase with ground through an object (e.g., a falling tree) of due to the proximity of two phases. The role of the protection mechanism is to quickly detect the situation, and disconnect ("clear") the faulted segment minimizing the impact to the grid. There are different mechanisms to detect and exercise action over the fault, and the teleprotection is fundamental in intersubstation telecommunication to maneuver the circuit breakers to interrupt continuity of the power line at its ends. From the telecommunications perspective, teleprotection (Figure 4.3) is the intersubstation communication between the relays.

IEC 61850-90-1:2010 "Use of IEC 61850 for the Communication Between Substations" gives details of different teleprotection use cases, among them, distance line protection with permissive overreach teleprotection scheme, distance line protection with blocking teleprotection scheme, directional comparison protection, transfer/direct tripping, interlocking, multiphase auto-reclosing application for parallel line systems, current differential line protection, and phase comparison protection. The current and voltage transformers' information is assumed as input to the working logic of the system.

The teleprotection equipment (IEC 60834-1:1999) is a very special telecommunication element, specially designed to be used along with a protection

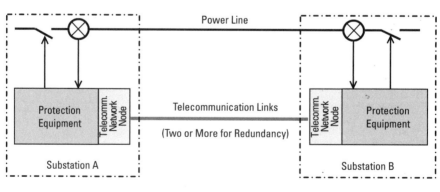

Current and Voltage Transformers
Circuit Breaker

Figure 4.3   Teleprotection and protection concept.

system to enhance its functionality. The teleprotection equipment works connected to a telecommunication link between the ends of the protected circuit, and transforms the information given by the protection equipment into a format that can be transmitted. Teleprotection devices send signals according to the different protection schemes.

The requirements that teleprotection service imposes over the telecommunication systems are the strictest ones in the utility domain. The data information transfer delay must not typically exceed 8 to 10 ms, which already rules out many telecommunication technologies. The use cases in IEC 61850-90-1 typically mention 5 ms, and in some cases 10 to 50 ms. The value depends on the voltage of the power line (the higher it is, the lower the allowed delay), and the constraint is given by the semi-period of the nominal grid frequency (50 or 60 Hz). IEC 60834-1:1999 establishes the performance requirements for teleprotection equipment and test methods for the different telecommunication media that can be used. The time it takes for the protection to act is critical, and depends not only on the telecommunication network but on the overall system (see Figure 4.4).

Two fundamental requirements of a protection system (including the teleprotection) are dependability and security. From a conceptual perspective, dependability evaluates the certainty that a protection system will operate when it is supposed to, while security evaluates the certainty that a protection system will not operate when it is not supposed to. From a formal perspective, IEC 60834-1:1999 gives definitions of both dependability and security based on probabilities, which are then translated into different performances of the protection schemes over different channel quality telecommunication systems.

Regarding the connection of protections with teleprotection equipment, IEEE C37.94-2002 describes the industry-standard interface for n x 64 kbps optical connectivity that the teleprotection device offers to the telecommunication network. This interface is compliant with ITU-T Recommendation G.704 and partially with G.775, but not with G.706. There are other standards still very much in use such as analog 4W E&M (4 Wires Ear and Mouth), RS-232, and ITU-T Recommendation G.703 codirectional 64 kbps.

4.2.3 Metering

A meter is an electricity measuring instrument which totalizes the electric energy consumed over a given time period. The meter and the associated metering services are an integral part of any utility service since they are necessary to charge customers based on electricity consumption. Meters can be found at different points through the grid; there are meters at customer premises, and there are meters at substations to control energy flow at grid interconnections.

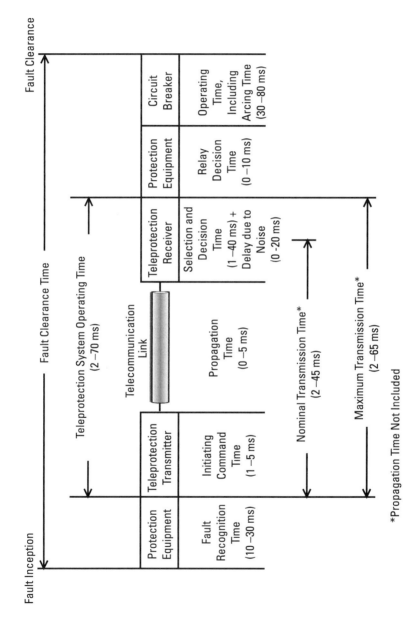

Figure 4.4 Typical times for teleprotection service (IEC 60834-1:1999).

Meters have traditionally been electromechanical devices, and the meters based on the Ferraris principle, patent US423210 from 1890 by Ottó Bláthy were eventually the most successful. Induction, "Ferraris" type, electromechanical meters have survived for more than a century.

From the 1970s, with the advent of the first analog and digital electronic circuits, electronic meters gained the attention of the industry (highly constrained by the low consumption, high reliability and accuracy needs of the meters themselves). In parallel, by applying telecommunications to meters, Automatic Meter Reading (AMR) systems started being used by utilities to remotely access meter readings.

Both ANSI and IEC have developed specifications for meters and metering services since the 1910s [6]. They specify accuracy classes for electromechanical and solid-state meters, and define associated tests. ANSI C12.1 is a well-known equipment performance standard for electricity meters. It is usually the reference for the performance and it influences specifications for electromechanical meters. This standard is similar in scope to a combination of IEC 62052-11:2003 and IEC 62053-11:2003. ANSI C12.20 specifies metering performance for 0.2% and 0.5% accuracy meters (class numbers represent the maximum percentage of metering error at normal loads [7]). The standard is similar to IEC 62053-22:2003. From a purely mechanical perspective, ANSI meters have a typically round form-factor, while IEC meters are square.

Apart from being electromechanical or electronic, meters can be classified according to their functions: the watt-hour meter measures energy (power over a period of time) and the watt meter measures power. There are also [8] multifunction meters, which measure both active and reactive power. Voltage ranging meters are able to adjust automatically to the voltage level of the input signal. Demand meters measure watts according to different periods of time, so as to be able to rate consumption that demands high peaks from the grid differently from those that have a more regular consumption (the same "average" consumption can create very different challenges –and need a higher investment– for the grid). Integrating demand meters measure the highest consumption in specific intervals (15, 30, or 60 minutes), and interval data meters measure power in predefined intervals (e.g., every 15 minutes for 24 hours).

The largest quantity of meters is with residential customers. They are typically located at the edge of the utility grid, inside customer premises. Different companies and markets have different ownership policies. Sometimes meters are owned by customers, or they may belong to the distribution company; in other cases there are third companies that run metering businesses separate from the DSO. In all cases the meters need to be read; this process has a cost as it implies access to customer premises.

Due to the interest to decrease costs, AMR systems began to be explored in the 1960s, in water, gas, and electricity utilities. In 1970 telephone lines were suggested for "telemetering" purposes. The examination of the patent evolution in this field shows the concerns for costs, availability of telecommunication technologies (e.g., telephone lines, PLC, radio), the scalability, the security, and the safety in AMR.

The first AMR systems could be classified between those accessing automatically meter readings from a central system and walk-by or drive-by type systems.

4.2.4   Video Surveillance and Other Security Needs

The protection of the substations from threats (internal or external; physical or cyber) is relevant because they are fundamental for reliable electricity delivery. Substations are very often unmanned and thus prone to any threat that could begin via a physical or cyber (remote) attack.

Traditionally, security in substations has been associated to the prevention of human intrusion (the intrusion of animals has also been studied, see IEEE 1264-2015). Intruders are unadverted people, thieves, vandals, disgruntled employees, or terrorists. A systematic analysis is always needed to assess the criticality of the substation, to evaluate its vulnerability (the weaknesses that may make the substation to get compromised) and finally to analyze risk scenarios and take, or do not take, actions.

The surveillance methods that, per se, do not require telecommunications include:

- Physical access prevention elements like fences and walls, gates and locks, barriers to energized equipment, and building design with appropriate materials;
- Hiding of assets (both substation and elements inside), signs and/or proper lighting of the property;
- Security patrols and site routine inspection.

On the side of the typical ICT-related systems that are used for security purposes, intrusion detection and access control systems are commonly found. Intrusion detection systems usually make use of all or some of the motion detection, the sound and video capabilities of sensors and ICT systems, to allow the detection and identification of the kind of threat. The information is sent to the surveillance control centers, with sound, image, and video of different qualities according to the capabilities of the telecommunications available.

In the sensors area, photoelectric, lasers, optical fibers and microwaves are used to detect intrusion. Door and windows alarm systems are also available to trigger alarms, both local (sirens, lights) and remote (on UCCs or directly to the police). Video cameras record actions from fixed positions, autonomously and/or remotely operated; the recording is often stored locally and partially transferred to the UCC when needed. However, video remote transfer needs to be considered very carefully, as it is usually the most bandwidth-demanding application. Video rates from 128 kbps to 2 Mbps can be affordable for important substations; if not, fixed images refreshed with enough frequency are a good alternative.

4.2.5   Intrasubstation Telecommunications

IEC 61850 series is the set of standards that configures the evolution of the communications inside the substation. The title uses the word "communications" as the series includes not only telecommunications, but also protocols and system aspects.

The IEC 61850 series is shown in Figure 4.5. It defines:

- The object models characterizing substation equipment (IED functions) and the information they exchange;
- A configuration language for SCADA operations and maintenance;
- The telecommunications connectivity inside the substation, made of two different bus types (see Figure 4.6):
  - Station bus: It is an OSI layer 2 Ethernet LAN. It provides connectivity for IEDs and other systems in the substation, including the IP routing element for SCADA master station external communication.
  - Process buses: Each bus with at least one layer 2 Ethernet switch, where the IEDs can connect among them and with the bay controllers. [See IEC 61850-5:2013, Communication networks and systems for power utility automation - Part 5: Communication requirements for functions and device model. Bay concept refers to the group of closely connected subpart of the substation with some common functionality (e.g., the switchgear between an incoming or outgoing line and the busbar). These subparts are called "bays" and managed by devices with the generic names "bay controller" and "bay protection" and their functionality represents an additional logical control inside the overall substation level, called "bay level".] The number of process buses varies depending on the number of equipment and its location.

The future evolution of intrasubstation communications will contemplate the existence of traditional RTUs and non-IEC 61850 elements for a period of time. Legacy serial connectivity supported by some RTUs will be transported

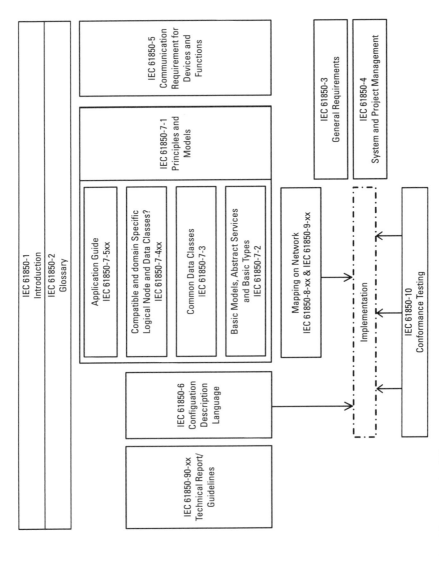

Figure 4.5 Relationship among the IEC 61850 components.

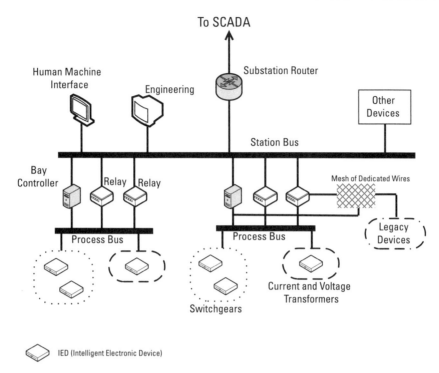

To SCADA

Human Machine Interface

Engineering

Substation Router

Other Devices

Station Bus

Bay Controller

Relay Relay

Mesh of Dedicated Wires

Process Bus

Process Bus

Legacy Devices

Switchgears

Current and Voltage Transformers

IED (Intelligent Electronic Device)

Figure 4.6 IEC 61850 architecture.

by the telecommunications infrastructure if encapsulated over the IP/Ethernet telecommunications infrastructure defined by the standard.

In telecommunication terms, the standard defines performance requirements for different message types. These performance requirements are fixed in terms of transfer times (see Table 4.1), time synchronization, and data integrity (measured in terms of residual error probability), and related to the protection functions of security, dependability and availability (see Section 4.2.2).

The implementation of telecommunications for IEC 61850 is based on high-bandwidth Ethernet LANs with a trend to incorporate:

- Optical fiber as the telecommunication medium, to avoid EMC issues from substation elements;
- Traffic prioritization mechanisms such as IEEE 802.1p (see IEEE 802.1D) for each type of data;
- Traffic isolation capabilities as defined in IEEE 802.1Q-2014;
- Reliability provided through immunity to electrical surges, electrostatic discharges, and other EMC aspects;

Table 4.1
IEC 61850 Telecommunication Transfer Time Classes

Transfer Time (ms)	Application Examples
>1,000	Files, events, log contents
1,000	Events, alarms
500	Operator commands
100	Slow automatic interactions
20	Fast automatic interactions
10	Releases, stats changes
3	Trips, blocking

- Improved service availability through the implementation of ring or meshed architectures, and mechanisms such as RSTP (IEEE 802.1w; see IEEE 802.1Q-2014);
- Security features to guarantee nonvulnerability of assets.

A noted element of this standard is the definition of generic object-oriented substation event (GOOSE) messages. These messages need to travel very fast within the local area network (LAN) in the substation, so that IEDs can react quickly to grid events. The delay of these messages is below 4 ms (a quarter of a period in a 60-Hz system), clearly a challenging requirement for any non-LAN telecommunication network. These messages can be mapped into the Ethernet and adapted to achieve the strict latency requirements.

IEC 61850 is expanding outside the substation physical domain [e.g., covering medium-voltage (MV) feeders to support both intersubstation communication and distributed generation control].

4.2.6 Mobile Workforce Telecommunications

Even in a world with technology and automation, human intervention is needed, from the UCC in charge of grid operations to crews deployed on field to intervene whenever manual operation is needed on-site.

Mobile telecommunications need to be provided for team coordination and control. Mobile telecommunications for utility crews have evolved along with technical progress and parallel to public commercial network services. The early adoption and advances on private mobile radio (PMR) systems were a consequence of public safety needs' adaptation (police, firefighters, blue-light services in general). Public mobile systems were widely adopted during the 1990s, and the first decade of the twenty-first century saw the expansion of data mobile telecommunications.

Traditional requirements of PMR systems have been different to the ones of public commercial mobile telecommunications. Although, for the latter, full-duplex transmission and high-frequency reuse were key, for the former, the adaptation to operational procedures and far-reaching coverage have been more important. Reference [9] included a list of requirements typical of PMR systems, for example, push-to-talk (PTT) operation mode with fast channel access. Other typical requirements are mobile-to-mobile telecommunications without base station support.

However, PMR systems have in many places been displaced by public mobile (cellular) systems. This trend has been favored by the good performance of these systems when under nonemergency conditions. Public systems, although rich in features, lack important aspects such as fast call establishment, user priority, infrastructure-related elements such as battery backup (when grid supply may fail) at base stations, and coverage in sparsely populated areas. Some of these issues can be solved from the utility perspective with the complement of land mobile satellite service. Commercially available satellite platforms provide a complementary coverage to that of mobile systems, and the incidents that affect both networks are generally uncorrelated. However, satellite services are usually expensive and require specialized terminals (often proprietary), and the coverage in urban areas and places where visibility of the satellite can be compromised is poor. So land mobile satellite service is often used as a backup for emergency situations.

Mobile telecommunication systems often support the normal operation of crews, driven by information systems that issue orders and programmed activities to them. Mobile crews access these systems in mobility situations when they are out of the office in their vehicles. It is normal to install access from these vehicles, or from the PCs and smart phones, to mobile commercial networks without mission-critical requirements to get access to corporate and operational applications and databases to drive and optimize their activity. Furthermore, positioning capacities [e.g., global positioning system (GPS)] help to organize the activities of different workforces spread over a territory.

4.2.7   Business (Corporate) Telecommunications

Many substations are still used as permanent utility offices. In these cases, the telecommunications infrastructure for corporate and operational needs overlap since staff has the needs of regular office employees (e.g., e-mail access, intranet access, ERP software access), as well as the need for operations-related applications (GIS access, operational databases access, and so forth). Telecommunication networks and services are needed to guarantee nonoperational and operational services access at these premises.

For the reason above, many utilities have telecommunication assets that have been developed and deployed over the years. The drivers for these investments were the operational needs and the lack of cost-effective public infrastructures before the 1990s. The global liberalization process of telecommunications was another stimulus for some utilities to build telecommunication infrastructures over existing utility poles and other rights of way. The telecommunication market demanded this infrastructure growth, as did shareholders. In particular, optical fiber cables were deployed on poles and inside ducts, which helped to develop a telecommunication network that utilities leveraged both for third-party needs (revenues) and provision of internal services (both operational and corporate). This infrastructure development took place with different business models: some utilities created telecommunication subsidiaries and outsourced services; others expanded the size and abilities of its telecommunications departments, preparing them for the smart grid challenges.

4.3   Smart Grid New Applications and Services

Telecommunication services associated to the smart grid evolve both because of new requirements in grid technologies, and because the electronics and telecommunication industries provide new possibilities for existing utility operations.

The evolution of telecommunication services for the smart grid largely focuses on the expansion of capacity (bandwidth), pervasiveness (number of end-points covered), reliability (telecommunications available when operations need to be made), and real-time nature. Additionally, new smart grid applications bring their own specific requirements.

4.3.1   Advanced Metering Infrastructure (AMI)

Electronics inside meters represents an opportunity for providing advanced metering services. While in the context of substation metering this is no revolution, in the context of residential metering, with meters present in all edges of the grid, the possibilities of ICTs are changing the landscape.

Electronic meters have advantages over electromechanical ones. A wide range of electrical parameters can be stored and accessed, a higher degree of accuracy can be obtained, the functionalities can be upgraded with the software/firmware, and the design is flexible enough to integrate telecommunication capabilities. When these electronic meters include telecommunication capabilities, with some of the characteristics described below, they may be called smart meters.

Advanced metering infrastructure (AMI) is an evolution of AMR. From the telecommunication perspective, AMI can be considered as such when two-way telecommunications and real-time features exist. From the functional perspective, measuring of a wide range of parameters (e.g., instantaneous voltage, current and power, cumulative energy consumption, and active and reactive components) is supported so that multiple time of use (TOU) tariff schemes are enabled remotely; remote connection/disconnection of supply points is also allowed. Smart meters are able to generate and communicate specific events and alarms.

AMI brings new possibilities. Utilities get advantages from all the new functionalities derived from getting real-time access to user consumption data (load patterns at grid edges). When plain AMR is combined with other options such as sending commands to a relay to remotely open and close the supply of energy for the customer, with features to get the information on which low-voltage (LV) feeder and phase meters connect to, and to quickly know that a group of customers is massively disconnecting from the grid ("last-gasp" meter functionality [10]), the utility has many possibilities of reducing costs and providing a better quality of service (QoS) over the distribution grid. Remote connection and disconnection greatly reduce the costs of physical access to the meter (e.g., new service provision, change of energy supplier, or bad debt). Feeder and phase identification maps every customer to the specific point of the grid where it is connected [11]. Fraud detection capabilities are also a clear advantage of AMI.

Customers can also directly get benefits from these new possibilities, as they can be made active participants of grid operation and status through time-based rate programs [12], usually referred to as dynamic pricing tariffs:

- TOU pricing: Different blocks of hours; the price for each period is predetermined and constant.
- Real-time pricing (RTP): Different pricing rates per hour.
- Variable peak pricing (VPP): Typically a mix of TOU and RTP for on-peak periods.
- Critical peak pricing (CPP): The price for electricity rises during specified periods of time when utilities observe or anticipate high energy prices or power system emergency conditions.
- Critical peak rebates (CPR): The customer gets refunds for any reduction in consumption relative to what the utility considered the customer was expected to consume, during specified periods of time when utilities observe or anticipate high energy prices or power system emergency conditions.

The term smart metering can be used as a synonym of AMI. Smart metering may imply a closer link with smart grids, as many utilities understand smart metering not only as an additional service of the smart grid, but as an opportunity to start building a smart grid. Smart metering then becomes, for these companies, the first stage towards smart grid. However, other utilities just refer to smart metering as the massive replacement of traditional meters with smart meters, be it a consequence of regulatory obligations or a genuine belief on its advantages.

Smart meters are characterized by a set of improved functionalities that exceed pure billing purposes and allow meters to behave as actual load control elements. In the last decades many intelligent or advanced meters have been deployed, but they do not share the same features. The European Commission established a common minimum set of functional requirements for electricity smart meters [13] as described below.

For customers, smart meters should:

- Provide readings directly to the customer and any third party designated by the consumer.
- Update the readings frequently enough to allow the information to be used to achieve energy savings. The consumer needs to see the information responding to their actions (the general consensus is that a minimum update rate of every 15 minutes is needed), as one of the objectives is to produce energy-saving incentives.

For metering operators, smart meters should:

- Allow remote reading of meters by the operator.
- Provide two-way communication between the smart metering system and external networks for maintenance and control of the system itself.
- Allow readings to be taken frequently enough for the information to be used for network planning.

For commercialization of the energy supply, smart meters should:

- Support advanced tariff systems: advance tariff structures, TOU registers, and remote tariff control, with the objective of helping consumers and network operators to achieve energy efficiencies and save costs by reducing the peaks in energy demand.
- Allow remote on/off control of the supply and/or flow or power limitation.

For security and data protection purposes, smart meters should:

• Provide secure data communications (privacy and data protection are required).
• Fraud prevention and detection, including access hacking.

For distributed generation (DG), smart meters should provide import/ export and reactive metering to allow renewable and local micro-generation data integration.

Another key component of a smart metering system is the meter data management system (MDMS). The MDMS can be defined as a system that collects and stores large quantities of meter data from a head end system and processes them into information that can be used by other applications such as billing, customer information systems, and OMSs [14]. The system needs to be able to evaluate the quality of the information and generated estimates if errors or gaps exist in the retrieved data [15].

The MDMS (Figure 4.7) is at the top of the architecture of the smart metering system. The intermediate layers may show different structure, depending on the capabilities of the telecommunication technology that provides the access.

The first architecture is the most traditional one. It includes the concept of the DC [16], as the collector of the meter information, and as a mediation device to any command that may be sent from the MDMS, or any event that may be originated in the smart meters. The location where the DCs are installed varies with the technology. If the technology is based on radio, the location of DCs can be flexibly chosen with the condition that the radio-enabled smart meters are within the reach of the system; if the technology is based on PLC, the location of DCs is usually at substations [primary substations (PSs) in the case of ultranarrowband PLC technology and secondary substations (SSs) if the technology is narrowband, PLC, or broadband power line (BPL)]. The convenience and cost of DC-based architectures is covered in [17].

The second typical architecture relies on the direct connection of the MDMS with the smart meters. Ideally, this connectivity would be based on IP, but this is not generally advisable with narrowband technologies, due to the capacity constraints. For this reason and because many times the upper layer protocols (and even the lower layer ones) are proprietary, the concept of the head end exists. This head end communicates with the MDMS, and either with the DCs or directly with the smart meters. The head end may exist in any of the architectures.

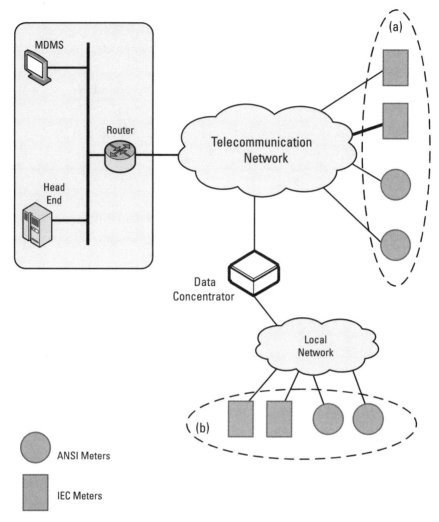

Figure 4.7   AMI architectures.

Different telecommunication technologies are used for smart meter access. Reference [18] included a summary; they can be basically classified as PLC and radio frequency (RF) ([19] focused on wireless).

Smart metering systems are usually also extended into the home of the customer to make them aware of their consumption; these data may come from the system or directly from the smart meter through the home area network (HAN). In-home displays and domestic appliances would connect to a HAN providing connectivity to the smart meter. This HAN may be realized with local wireless or wireline means (typically RF or PLC). However, if the information from the smart meter is being obtained from the MDMS, the basic connectivity would be through any Internet access connectivity.

At application layer, apart from the multiple proprietary options, the two more widely accepted standards are the ones specified by ANSI (C12.22 and C12.19) and IEC (62056 series). ANSI C12.22 specifies the communication between a smart meter and the MDMS. ANSI C12.19 defines the data structures that are to be transported by the ANSI C12.22. A similar structure can be seen in the IEC 62056 series, which basically covers the DLMS/COSEM suite [20]. The two different application layers above support different PHY and MAC technologies.

The choice of an AMI/smart metering telecommunication infrastructure is a key one for the evolution of the smart grid in the utility. Smart metering may be seen as the foundation for the overall grid modernization within the utility long-term vision. The telecommunication system deployed for it, if seen as part of that smart grid evolution, will need to accommodate and anticipate the future needs, to have the flexibility to handle different applications, being smart metering just one of them [21].

4.3.2 Distribution Automation (DA)

DA is the automation of utility assets that lay on distribution grids outside substations. It can be considered to range from just an extensive deployment of traditional automation capabilities of the distribution grid [22], to the automation of all the functions related to the distribution system from information collected from the grid elements (this would include distribution assets and smart meters) [23]. The broadest scope of DA may include both remote control functions (i.e., SCADA) and information from AMI/smart metering systems, as DMSs will use the information retrieved from the smart meters as well. However, in this section DA will just refer to the part of the distribution grid between the PSs and the customer loads where the smart meters are installed, but excluding them. It includes some of the functionalities of what it is also known in literature as advanced distribution automation (ADA) involving the recent progresses in automation out of the substation.

DA is an extension of the automation functions that began in the substation, further down into MV feeders, and potentially will reach to the consumers. DA is part of the active grid concept where the grid dynamically auto-configures to optimize its performance. The concept is also closely related to the target of having a grid that self-diagnoses and self-heals and meets the long-term objective of including the distribution grid under automatic protection schemes to protect and switch (detect, isolate, and attempt to restore the maximum of the grid) where and when convenient, adapting to variable network conditions. Telecommunications-originated protection techniques could be taken as a reference [22] in the smart grid.

The metrics (indices) used to measure grid performance and reliability can be found in IEEE 1366-2012. The different indices are classified as "sustained interruption," "load based," and "momentary interruption"; the first two ones measure the unavailability of the service based on the customers affected, and the third one is based on the load interrupted.

Some commonly used indices are:

- System Average Interruption Frequency Index (SAIFI): This index indicates how often the average customer experiences a sustained interruption over a predefined period of time.

- System Average Interruption Duration Index (SAIDI): This index indicates the total duration of interruption for the average customer during a predefined period of time. It is commonly measured in minutes or hours of interruption.

- Customer Average Interruption Duration Index (CAIDI): This index represents the average time required to restore service.

- Customer Average Interruption Frequency Index (CAIFI): This index gives the average frequency of sustained interruptions for those customers experiencing sustained interruptions. The customer is counted once, regardless of the number of times interrupted for this calculation.

- Momentary Average Interruption Frequency Index (MAIFI): This index indicates the average frequency of momentary interruptions.

DA functional grid elements are tools to improve the grid performance:

- Reclosers: From an electrical perspective, a recloser is functionally similar to a circuit breaker used to open a circuit when there is a failure. However, the recloser is capable of monitoring the status of a feeder to break it if a fault is detected. It includes the function of automatically trying to reconnect after a period of time, assuming that the fault could be a transitory condition. After a predetermined number of reconnection attempts, the breaker "locks out" and keeps open and is just closed through remote or manual intervention.

- Switches: A switch is used to sectionalize sections of the grid, when electricity needs to be routed (e.g., in the case of a fault).

- Capacitor banks: These elements are used to reduce the reactive current flowing through the grid, to keep the power factor close to 1. Although these elements can be deployed at the substations, they perform better when they are closer to the loads, and this is why they can be placed on the feeders. Real-time measurements indicate how many capacitors need to be connected.

- Distribution transformers: Transformers vary in size, disposition, and capacity depending on the grid type (e.g., United States versus Europe) and area (urban versus rural). The number of supply points ranges from just a few customers to hundreds of them. Although voltage, current, and temperature are not typically measured, these magnitudes are important not only to balance the loads in the grid but to control the lifetime of the transformers.

DA requires inclusion of telecommunications for these elements to enable their remote operation.

4.3.3 Demand Response (DR)

Demand-side management (DSM), which has been promoted actively by the EPRI since the 1980s, includes all the activities performed by the utilities to influence the customer demand to balance the grid electricity supply and the demand. These activities are quite heterogeneous, and, apart from the energy efficiency initiatives, the most important one is DR. DR is considered sometimes interchangeable with DSM, although it just refers to a particular area of DSM where the focus is on the ability, with the help of customers, of shifting loads in a reactive manner, with the help of two-way telecommunications on the consumer side.

In the broader context, DSM includes all the activities designed to influence customer use of electricity to produce the desired changes in the load shape (pattern and magnitude) (see Figure 4.8). DSM is usually associated with long-term and medium-term changes; DR is focused on the short-term needs.

DR actions can be classified according to the time-scale. Different DR actions can be classified according to the time that it takes to plan its implementation and the time that it will take to get its effects effectively on the grid.

DR involves all types of consumers. The programs for each group have different telecommunication requirements. Residential environments are the most challenging due to the massive amount of end-points and the difficulty to provide access to most or all of them.

DR initiatives are also classified according to the need of telecommunications. System-wide planning and operational management of existing resources do not need telecommunications and need a long time to be planned and im-

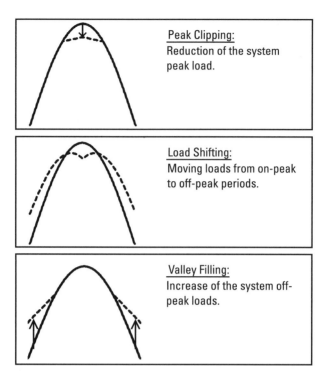

Figure 4.8   Load shape management [24].

plemented. Direct load control and automated DR are examples of initiatives needing telecommunications and delivering short-term results.

Direct load control is used with domestic appliances and refers to the way in which the utility can directly manage consumers' appliances (e.g., heat pumps, water heaters, and clothes dryers) to impact on the total load of selected grid areas. Consumers sign agreements with service providers to let them decide when the appliance is to be connected or disconnected. The user has generally the capability to override the orders from service providers under preestablished conditions. Direct load control can be implemented with different technologies, and sometimes just one-way telecommunication is needed to control the appliances inside customer premises. One such technology is ripple control, described in [25] where voltage pulses in specific (audio) frequencies are injected into the MV grid, so they propagate up to LV loads, and these pulses encode control messages (receiver requirements are specified in IEC 62054-11:2004).

Automated demand response is a more sophisticated, closer to real-time control of home appliances and other home loads (e.g., lighting). The utility, through a modern telecommunication network and a proper application-layer language (e.g., OpenADR [26]), establishes a dialogue with home appliances to control consumption with a high granularity.

4.3.4   Distributed Generation (DG)

DG is defined within the concept of distributed energy resources (DERs) including power sources smaller than traditional centralized generation plants, and energy storage, in close proximity to customers.

The reference to DG as one of the smart grid services reflects a growing reality, present in grid locations where traditionally only loads (supply points) were present. DG refers to power sources that can be connected to the grid at any point (MV or LV grid), thus reverting the traditional, one-way flow of energy from generation to consumers through the transmission and distribution infrastructure. The most popular distributed power sources are wind energy turbines (producing from 1 MW to hundreds of megawatts in farms), solar power (photovoltaic plants; from 2.5 kW of a roof-top panel, and up to several megawatts of power in big plants), fuel cells (up to 100 kW), biomass/biogas, and even electric vehicles when injecting power from their batteries into the grid.

DG sometimes generates directly alternating current (ac) (e.g., wind turbines) or direct current (dc) (e.g., solar power). The challenges are different in each case. If power is produced as ac, the DG ac source needs to be synchronized with the grid in order to be connected. In the case of dc, a dc-ac conversion is needed through power inverters. The harmonics created by them (above 10 kHz, and as far as 150 kHz) can be harmful for certain PLC technologies. Battery storage is needed at the sources where a constant supply of energy cannot be guaranteed (e.g., solar and wind).

DG is not a service that the utilities as such need to deliver, but a new grid element to be controlled. DG can have a negative effect in the stability and operation of the grid; system balance (adjusting the system to fluctuating power requirements) must be achieved while, at the same time, situations such as unintentional islanding (the situation where a DG power sources feeds a segment that is considered out of service by the utility) must be avoided. Telecommunications are fundamental to support the proper integration of DG sources. For grid stability purposes, DG must be remotely monitored in terms of voltage level and power quality (frequency, flicker, power factor, harmonics); voltage stability becomes difficult if active voltage control is not in place. For safety reasons, DG sources should be able to get disconnected from the grid both by itself and with a remote command.

The development of telecommunication means may be a challenge for individual users producing energy in areas where the access to the telecommunication service provider's services is not available. It is fundamental that DG is connected to the grid wisely, ensuring access to telecommunication networks which render the DG investment operational from both safety and grid stability perspectives.

4.3.5   HAN Domain

The engagement of consumers in the energy market will also mean that they take responsibility for the part of the grid under their ownership. Although countries present slightly different situations, utilities' responsibilities and ownership of different grid segments are fixed by regulation. The HAN is present in the customers' properties.

DR programs sometimes involve in-home appliances that might be left to the control and management of the utility. This possibility hinges on the availability of any form of telecommunications connectivity with customer premises. If this in-home or in-building connectivity is generalized, HANs may connect all appliances within the home or building, so that not only the utility but also the customer profits from these elements communicating among them to optimize energy efficiency.

Systems developed to achieve the goals above are called home energy management system (HEMS) and building energy management system (BEMS) (HEMS will be used to simplify the use of acronyms). HEMS is a comprehensive set including any product or service which monitors, controls, or analyses energy in the home. Figure 4.9 shows the objectives and requirements of HEMS within the smart grid. It may include DR programs, home/building automation services, energy management, and integration of DG.

The challenge of HEMSs is not with the telecommunication technologies involved, but on the information technology (IT) systems to take advantage of energy efficiency opportunities, together with the possibility of integration with utility systems to participate in grid energy management.

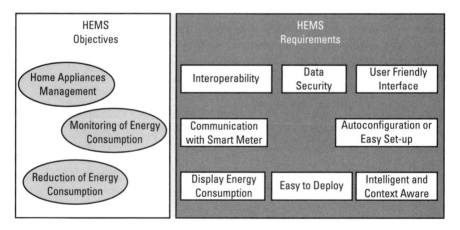

Figure 4.9   The objectives and requirements of HEMS within the smart grid [27].

4.3.6 Cybersecurity

Smart grid brings new security concerns, further than the traditional prevention of intrusion or physical attacks. Cyberattacks become a reality in this new context as ICTs become an integral part of the operational processes of the grid (see the vision of IEEE 1402-2000).

Cybersecurity is, according to the ITU-T Recommendation X.1205, "the collection of tools, policies, security concepts, security safeguards, guidelines, risk management approaches, actions, training, best practices, assurance and technologies that can be used to protect the cyber environment and organization and user's assets. Organization and user's assets include connected computing devices, personnel, infrastructure, applications, services, telecommunications systems, and the totality of transmitted and/or stored information in the cyber environment." The ITU-T X series of Recommendations deals with data networks, open system communications and security.

The term cybersecurity is relatively new. IEEE 1402-2000 referred to it as "electronic security," in avoidance of substation entry "via telephone lines or other electronic-based media for the manipulation or disturbance of electronic devices." The electronic security mechanisms that were foreseen then were those of computer security systems, further dividing it into identification, authentication, and auditing. Among the typical security measures mentioned, there are those related to passwords, dial-back verification in the access through modems, virus scanning, encrypting, and encoding. Any recent analysis shows that this guide is not state-of-the-art anymore. Two-way broadband telecommunications, the hyper-abundance of electronics at substations, the use of public networks, the complexity of the architectures and operational procedures have forced evolution in network cybersecurity standards. A complete reference can be found in [28].

Smart grid cybersecurity is similar to generic IT cybersecurity, but there are some specifics to it. Focusing on substations, where most IEDs are placed, some particular aspects are:

- Security is an eclectic discipline, and it must be understood as a process in constant evolution.
- Networking for smart grid is different from networking for pure IT purposes and networking security needs to be specifically adapted to the smart grid. Firewalls for substation access evolve into intrusion detection system (IDS), intrusion prevention system (IPS), and eventually including deep packet inspection (DPI) capabilities to examine the application data payload.
- Private and isolated networks are not intrinsically secure.

- Smart grid protocols, although not as publicly known as others, do not offer extra "security by obscurity": there are always motivated attackers, some of them coming from the utility industry, who challenge that assumption.
- Virtual private networks (VPNs) over public networks offer different security characteristics. VPNs, if encrypted, do not allow DPI to perform their function.

Some of the challenges of cybersecurity for smart grids are:

- The long-term nature of utility investments does not favor the continuous introduction of new security measures as existing hardware may not support a new secure software (e.g., smart meters may not support advanced encryption and may need to be replaced).
- For the reason above, in those systems where legacy operating systems are present, it is likely that some known vulnerabilities exist, until the hardware is replaced.
- Open protocols and the availability of hardware (test devices) or software developed for generalist hardware platforms provide tools for attackers to experiment with.
- Smart grid is still in its infancy in many aspects; there is no comprehensive (technical or organizational) commonly agreed approach to cybersecurity by utilities.
- The large amount and widespread deployment of electronics connected to smart grid systems make the smart grid cybersecurity challenge an unprecedented one. Potentially, relevant assets for the operation of a critical system could be attacked from anywhere.

There are measurements that can be taken to minimize the probability of cyberattacks: control of network traffic, control of the electronic devices themselves removing all default-configuration unnecessary features and disabling unused interfaces; password and authentication policies; regular drills; encryption of protocols; auditing of private and third parties' telecommunication networks. However, an intrusion may eventually happen, so it needs to be detected. To this end, anomalies in the systems need to be constantly studied and analyzed to detect changes in normal behavior. In the case that an intrusion is detected, the response needs to consist of recording, reporting, and restoring [29]. Recording allows further study of what happened to avoid it in the future; reporting is fundamental to limit the impact of any intruder; and restoring needs to provide the desired availability of the system.

4.3.7 Other Smart Grid Elements

There are other smart grid-related elements worthwhile mentioning. The smart grid concept is constantly evolving.

4.3.7.1 Distributed Storage (DS)

The principle that electricity cannot be stored on the grid is being revisited with the progress of technology. The structure of the grid was designed with this assumption, and many of the operational activities are designed based on it (i.e., matching production with demand). In the past, strategies such as pumped-storage hydro served to use some extra generation capacity. However, the process itself consumed energy and was not the most efficient way of storage.

Storage is always based on the transformation of electricity into a different type of energy. The evolution of storage systems focuses first on how to ensure efficiency in the conversion; the rest of the elements to consider are time response (i.e., how long it takes to have the load available), power rating, and discharge time. The continuous availability of larger capacity, faster-charging, and more efficient batteries and other storage elements (flywheels and super-capacitors) is creating new possibilities for DS to support DG, electric vehicles (EVs), and other applications.

4.3.7.2 Electric Vehicles (EVs)

Within the context of fossil fuel scarcity and the target to reduce pollution, EVs come up as new elements that will make the smart grid evolve. The effect of the massive introduction of EVs will manifest itself in the need to reinforce the distribution grid at points where these vehicles will be charged (pervasive); also, these EVs will also be used as mobile DSs if charged batteries are used to supply energy to the grid.

4.3.7.3 Microgrids

The concept of microgrid represents a group of consumers connected to a small grid with a generation source capable of remaining energized when disconnected from the utility grid. The group may be neighbors in a building, a campus, or a community that is connected to the utility distribution grid as well.

Microgrids are not specific to smart grids, but a special use-case that comes with the generalization of DG, DS, and ICTs. In fact, many of these microgrids started using generation sources (e.g., fuel generators) intended for outage situations or to provide extra power to the microgrid. This extra power can also be injected to the utility grid in case of need. The challenge is then to manage a multiplicity of microgrids which require smart integration, similar to the DG scenario.

Microgrid energy management system (MEMS) is the concept that has been defined to coordinate energy transactions with the utilities' DMS and EMS (for energy management, retail markets, and DR purposes).

Two IEEE references will be mentioned. The IEEE 1547.4-2011 standard "Guide for Design, Operation, and Integration of Distributed Resource Island Systems with Electric Power Systems" defined the basic microgrid connection and disconnection process for a stable and secure electrical interconnection of distributed energy resources to the distribution grid. The IEEE P2030.7 project, "Standard for the Specification of Microgrid Controllers," is directly related to the microgrid operation with the MEMS and includes the control functions that define the microgrid as a system that can operate autonomously or be grid-connected and seamlessly connect to and disconnect from the main distribution grid. The project addresses the common operational technical issues associated to the MEMS, independently of the topology, configuration, or jurisdiction of the microgrid, and presents the control approaches required from the DSO and the microgrid operator.

4.3.7.4   Transmission Grid Technologies

There are applications that are specific to transmission grids, as a consequence of the high value of transmission assets, the critical nature of the transmission grid, and the lower number of assets involved. However, the trend is for transmission-specific applications to progressively be used in distribution grids as well.

Dynamic Line Rating (DLR)

Transmission system limits are imposed by the capacity of power lines [30]. Transmission system operator (TSOs) usually make static calculations for their transmission lines, estimating the maximum amount of current that the power line's conductors can carry under certain, generally conservative, climatic conditions; these calculations establish a worst-case scenario that prevents the use of the lines to their limits when the existing conditions are not the ones considered.

One alternative to the worst-case static calculation is to adjust ratings more frequently; DLR technologies enable TSOs to determine capacity and apply line ratings in real time, to use the extra capacity when it is available. DLR avoids unnecessary and costly investments.

A DLR system is basically comprised of IEDs (with sensors to measure wind speed, ambient temperature, solar radiation, and even parameters from the power line) that communicate with the EMS to support algorithms that obtain the allowable rating for each specific conductor and environmental conditions. These ratings can be incorporated into a control system, such as a SCADA system or EMS.

FACTS (Flexible AC Transmission System)

Due to the nature of the loads in the grid, transmission systems have higher losses in the presence of reactive loads. These loads imply not only higher losses, but also voltage drops in the power lines.

Utilities take actions to compensate the reactive component of the loads and install capacitor banks either in parallel or in series with the power lines. FACTS Controllers [31] have been developed to control reactance through power electronics (thyristors). FACTS introduce controllability of power in the grid and enhance the usable capacity of power lines.

Synchrophasors in Wide Area Measurement System (WAMS)

Synchrophasors are phasor measurement units (PMUs) with time-stamps that, when referred to a common time reference, can follow the exact synchronization of the voltages and currents in all grid positions where they are installed. PMUs collect voltage and current values (amplitude and phase) with a frequency related to the nominal power frequency, so the grid can be characterized (10, 25, 50, and 100 measurements per second for 50-Hz systems; 12, 15, 30, 60, and 120 for 60-Hz systems). IEEE C37.118.1-2011, IEEE C37.118.1a-2014, amendment 1, and IEEE C37.118.2-2011, specify measurements and data transfer formats, respectively.

Synchrophasors are incorporated in a WAMS. PMUs, with time-stamped measurements, report to a system that uses the information to monitor the overall status of the grid with the ultimate objective of preventing blackouts in large-scale interconnected grids.

From all the initiatives to study synchrophasors worldwide [32], North American SynchroPhasor Initiative (NASPI) [33] in the United States is especially relevant, as it is producing an architectural reference. The architecture for the WAMS is based on a data bus connecting the phasor gateways that get connected to the PDCs from each utility; these PDCs receive the PMUs with time stamps from the selected locations of the grid. The architecture is very specific in the availability requirements and latencies for the different data classes: the feedback control is specified with an availability of 99.9999% and latency below 50 ms; the feedforward control is defined with an availability of 99.999% and latency below 100 ms.

References

[1]   Godfrey, T., *Guidebook for Advanced Distribution Automation Communications—2015 Edition: An EPRI Information and Communications Technology Report*, Palo Alto, CA:

Electric Power Research Institute, 2015. Accessed March 22, 2016. http://www.epri.com/abstracts/Pages/ProductAbstract.aspx?ProductId=000000003002003021.

[2]   Telemetry & Remote SCADA Solutions, *SCADA Systems*, Ontario: Schneider Electric, 2012. Accessed March 22, 2016. http://www.schneider-electric.com/solutions/ww/en/med/20340568/application/pdf/1485_se-whitepaper-letter-scadaoverview-v005.pdf.

[3]   Lee Smith, H., "A Brief History of Electric Utility Automation Systems," *Electric Energy T&D Magazine*, Vol. 14, No. 3, April 2010, pp. 39–46.

[4]   DNP Users Group. "About DNP Users Group," 2011. Accessed March 22, 2016. https://www.dnp.org/Pages/AboutUsersGroup.aspx.

[5]   Sato, T., et al., *Smart Grid Standards: Specifications, Requirements, and Technologies*, New York: Wiley, 2015.

[6]   "A Comparative Introduction to ANSI Metering Standards," *Metering & Smart Energy International*, June 30, 2003. Accessed March 22, 2016. https://www.metering.com/a-comparative-introduction-to-ansi-metering-standards/.

[7]   Seal, B., and M. McGranaghan, *Accuracy of Digital Electricity Meters: An EPRI White Paper*, Palo Alto, CA: Electric Power Research Institute, 2010. Accessed March 22, 2016. http://www.epri.com/abstracts/pages/productabstract.aspx?ProductID=000000000001020908.

[8]   Grigsby, L. L., *Power Systems*, 3rd ed., Boca Raton, FL: CRC Press, 2012.

[9]   Gray, D., *Tetra: the Advocate's Handbook: From Paper Promise to Reality*, TETRA Advocate, 2003.

[10]   Energy Networks Association (ENA), *Analysis of Network Benefits from Smart Meter Message Flows*, 2012. Accessed March 22, 2016. http://www.energynetworks.org/assets/files/electricity/futures/smart_meters/Network%20benefits%20of%20smart%20meter%20message%20flows%20V1%200%200300312.pdf.

[11]   Sendin, A., et al., "Enhanced Operation of Electricity Distribution Grids Through Smart Metering PLC Network Monitoring, Analysis and Grid Conditioning," *Energies*, Vol. 6, No. 1, 2013, pp. 539–556.

[12]   SmartGrid.gov, "Time Based Rate Programs." Accessed March 23, 2016. https://www.smartgrid.gov/recovery_act/time_based_rate_programs.html.

[13]   European Commission, *Commission Recommendation of 9 March 2012 on preparations for the roll-out of smart metering systems (2012/148/EU)*," Brussels, 2012.

[14]   Matheson, D., C. Jing, and F. Monforte, "Meter Data Management for the Electricity Market," *International Conference on Probabilistic Methods Applied to Power Systems*, Ames, IA, September 13–16, 2004, pp. 118–122.

[15]   Moore, S., *Key Features of Meter Data Management Systems*, Liberty Lake, WA: Itron, 2008. Accessed March 23, 2016. https://www.itron.com/PublishedContent/Key%20MDM%20Features%20Whitepaper_FINAL.pdf.

[16]   *Smart Grid & Energy Solutions Guide*, Dallas, TX: Texas Instruments, 2015. Accessed March 23, 2016. http://www.ti.com/lit/sl/slym071o/slym071o.pdf.

[17] Zhou, J., R. Qingyang Hu, and Y. Qian, "Scalable Distributed Communication Architectures to Support Advanced Metering Infrastructure in Smart Grid," *IEEE Transactions on Parallel and Distributed Systems*, Vol. 23, No. 9, 2012, pp. 1632–1642.

[18] Gungor, V. C., et al., "Smart Grid Technologies: Communication Technologies and Standards," *IEEE Transactions on Industrial Informatics*, Vol. 7, No. 4, 2011, pp. 529–539.

[19] Meng, W., R. Ma, and H. H. Chen, "Smart Grid Neighborhood Area Networks: A Survey," *IEEE Network*, Vol. 28, No. 1, 2014, pp. 24–32.

[20] DLMS User Association, "What Is DLMS/COSEM." Accessed March 23, 2016. http://www.dlms.com/information/whatisdlmscosem/index.html.

[21] The Modern Grid Strategy, *Advanced Metering Infrastructure*, Morgantown, WV: National Energy Technology Laboratory, 2008. Accessed March 23, 2016. https://www.netl.doe.gov/File%20Library/research/energy%20efficiency/smart%20grid/whitepapers/AMI-White-paper-final-021108--2--APPROVED_2008_02_12.pdf.

[22] Bush, S. F., *Smart Grid: Communication-Enabled Intelligence for the Electric Power Grid*, New York: Wiley-IEEE Press, 2014.

[23] Budka, K. C., J. G. Deshpande, and M. Thottan, *Communication Networks for Smart Grids. Making Smart Grid Real*, London: Springer-Verlag, 2014.

[24] Gellings, C. W., *The Smart Grid: Enabling Energy Efficiency and Demand Response*, Lilburn, GA: The Fairmont Press, 2009.

[25] "Load Management and Ripple Control," in *Switchgear Manual*, 12th ed., S. Kaempfer and G. Kopatsch, (eds.), Mannheim: ABB AG, 2012, pp. 769–770.

[26] OpenADR Alliance, "Overview." Accessed March 23, 2016. http://www.openadr.org/about-us.

[27] Rossello-Busquet, A., and J. Soler, "Towards Efficient Energy Management: Defining HEMS and Smart Grid Objectives," *International Journal on Advances in Telecommunications*, Vol. 4, No. 3 & 4, 2011, pp. 249–263.

[28] The Smart Grid Interoperability Panel – Smart Grid Cybersecurity Committee, *NISTIR 7628 Revision 1 - Guidelines for Smart Grid Cybersecurity*, Gaithersburg, MD: National Institute of Standards and Technology, 2014.

[29] McDonald, J. D., *Electric Power Substations Engineering*, 3rd ed., Boca Raton, FL: CRC Press, 2012.

[30] Energy Sector Planning and Analysis (ESPA), *Dynamic Line Rating Systems for Transmission Lines - Topical Report*," U.S. Department of Energy – Electricity Delivery & Energy Reliability, 2014. Accessed March 23, 2016. https://www.smartgrid.gov/files/SGDP_Transmission_DLR_Topical_Report_04-25-14_FINAL.pdf.

[31] Hingorani, N. G., and L. Gyugyi, *Understanding FACTS: Concepts and Technology of Flexible AC Transmission Systems*, New York: Wiley-IEEE Press, 2000.

[32] Phadke, A. G., "The Wide World of Wide-Area Measurement," *IEEE Power and Energy Magazine,* Vol. 6, No. 5, 2008, pp. 52–65.

[33] North American SynchroPhasor Initiative, "Vision and Mission Statements," 2015. Accessed March 23, 2016. https://www.naspi.org/vision.

5

Telecommunication Technologies for Smart Grid Networks

Telecommunications have become a needed commodity and are widely used in residential, commercial, and industrial environments. Different technologies have been developed to support applications in various environments. The evolution of these technologies adapts them to changing end-user needs, producing upgrades to existing technologies that coexist with disruptive novelties. Usually all of them coexist until investments get amortized and the electronics in the designs can no longer be maintained.

This chapter focuses on the telecommunication technologies considered useful for smart grid applications. Each technology will be considered independently, although in real smart grid implementations, technologies are combined within the telecommunication systems.

5.1 Transport Technologies

Transport refers to the point-to-point bidirectional transmission of data within a telecommunication network. Transport technologies are usually considered for backbone connectivity in the core and access blocks of the network (see Chapter 6). Transport technologies are often associated with high bandwidths, because they support the aggregation of multiple services.

5.1.1 Time Division Multiplexing (TDM)

TDM is a technology that allows multiplexing several signals over the same physical channel, assigning different time slots for transmission of each logical channel. In TDM, time is the resource to be shared by logical channels that use all the available bandwidth when transmitting, whereas in frequency division multiplexing (FDM), frequency is the resource shared by logical channels that transmit continuously when they have a certain frequency band assigned to them. TDM is used for the transport of different services organized in the time domain at a specified frequency with a predefined structure over a physical medium.

The consolidation of TDM came with its use in the digital telecommunications, first to multiplex digitized voice channels, and eventually any kind of data. In the example of digitized voice, channels are coded in 56 or 64 kbps channels (depending on the standard) and transported over a digital hierarchy (Figure 5.1).

TDM evolved in the various standardization efforts worldwide into different versions of what is generally called plesiochronous digital hierarchy (PDH). E-Carrier, T-Carrier, and J-Carrier [1] are the main TDM technologies implemented in Europe, the United States, and Japan, respectively (some common data rates are shown in Table 5.1). Higher data rate interfaces multiplex lower data rate interfaces (e.g., in E-Carrier, thirty-two 64 kbps channels are multiplexed in an E1-2,048 Mbps; some of these channels are used for signaling).

PDH is standardized, among others, in ITU-T Recommendations G.704 "Synchronous Frame Structures Used at 1544, 6312, 2048, 8448 and 44736 kbit/s Hierarchical Levels," and G.732 "Characteristics of Primary PCM

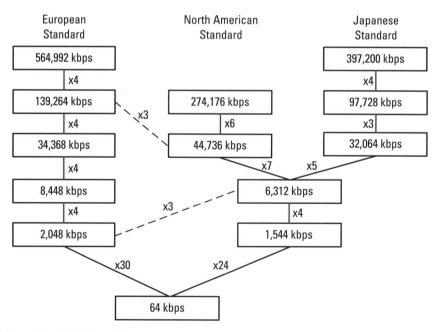

Figure 5.1 TDM hierarchies.

Table 5.1
TDM Technologies Data Rates

ITU	ATIS (ANSI)	Data Rate (kbps)
	T1	1,544
E1		2,048
	T2	6,312
E2		8,448
E3		34,368
	T3	44,736
E4		139,264

Multiplex Equipment Operating at 2048 kbit/s." PDH is also specified by several ATIS standards, such as ATIS 0600403 "Network and Customer Installation Interfaces - DS1 Electrical Interface" (formerly ANSI T1.403) and ATIS 0600107 "Digital Hierarchy - Formats and Specifications" (formerly ANSI T1.107).

5.1.2 SDH/SONET

Synchronous digital hierarchy (SDH) and synchronous optical network (SONET) are transport protocols defined as an evolution of TDM technologies with the purpose of increasing transmission capacities. SDH is standardized by ITU-T Recommendations G.803 "Architecture of Transport Networks Based on the Synchronous Digital Hierarchy (SDH)" and G.783 "Characteristics of Synchronous Digital Hierarchy (SDH) Equipment Functional Blocks," among others. SONET is formalized in several standards, such as ATIS 0900105 "Synchronous Optical Network (SONET) – Basic Description Including Multiplex Structure, Rates, and Formats" (formerly ANSI T1.105) and ATIS 0600416 "Network to Customer Installation Interfaces - Synchronous Optical NETwork (SONET) Physical Layer Specification: Common Criteria" (formerly ANSI T1.416). SONET is used in North America and Japan, while SDH is typically used elsewhere.

The advantages [2] of SDH/SONET in comparison with TDM include simplified add/drop functions, higher availability, overhead reduction and traffic encapsulation capabilities. Both protocols are defined to be used over optical fiber channels, but also define electrical interfaces (e.g., to be connected to microwave radio links).

The hierarchy of interfaces defined by both standards is provided in Table 5.2. To encapsulate traffic from different sources, a hierarchy of transport modules (VC in SDH and VT in SONET) is also defined (Figure 5.2). SDH/SONET is able to transport, for example, PDH, IP/Ethernet, and legacy asynchronous transfer mode (ATM) signals (Figure 5.3).

There are four types of devices in the SDH/SONET architecture:

Table 5.2
SDH/SONET Interfaces

SDH	SONET	Bit Rate (Mbps)
STM-0	STS-1	51.84
STM-1	STS-3	155.52
STM-4	STS-12	622.08
STM-16	STS-48	2,488.32
STM-64	STS-192	9,953.28
STM-256	STS-768	39,813.12

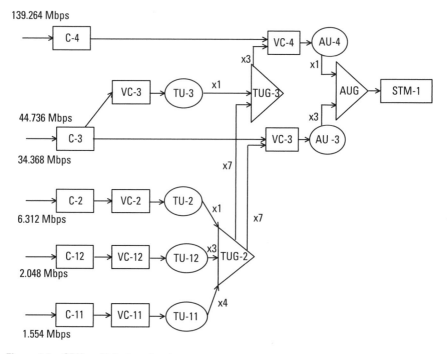

Figure 5.2   SDH multiplexing structure.

- Regenerator: It is used to regenerate the clock and the amplitude of the signal.

- Terminal multiplexer: It combines several channels (e.g., PDH signals) into a higher-rate SDH signal.

- Add-drop multiplexer (ADM): It is used to extract or insert transmission units (called tributaries) into the main channel (called aggregate).

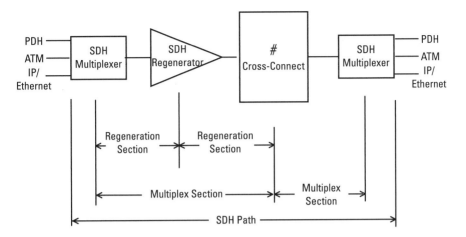

Figure 5.3 SDH path.

- Cross-connect: It is a connection matrix that, conceptually similar to a routing element, is used to connect any channel at the input to any channel at the output.

5.1.3 Wavelength Division Multiplexing (WDM)

WDM represents the next step in terms of increasing transport capacity. WDM is a technology that multiplexes several optical signals in one single pair of optical fibers (as an exception, one single fiber) using a different wavelength for each of the transported signals. It is the same as the FDM concept, but refers to wavelength, another way to represent frequency when it is extremely high.

The main advantage of this technology is the expansion of the capacity of the network's optical fiber cables without new investments in infrastructure. Thus, the same existing network is able to multiply the available bandwidth. Modern systems can multiply the capacity of each pair of fibers to more than 30 Tbps covering distances of hundreds of kilometers [3].

WDM can encapsulate any kind of traffic coming from other technologies, multiplexing it at the origin and demultiplexing at the destination. This makes WDM transparent for already existing devices such as SDH/SONET nodes as shown in Figure 5.4.

WDM is presented in two different options, coarse and dense WDM (CWDM and DWDM, respectively), classified by capacity and cost (higher in DWDM). Both are based on the same concept but differ in the number of channels (wavelengths) used and the spacing between two consecutive channels, as well as the ability to amplify them to reach longer distances.

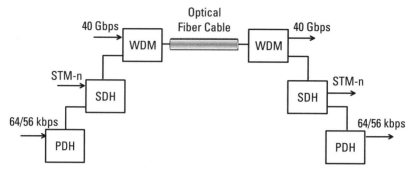

Figure 5.4 Example of optical fiber reuse, with the aggregation of different technologies.

ITU-T standardizes WDM in several Recommendations, such as G.694.1 "Spectral Grids for WDM Applications: DWDM Frequency Grid" and G.698.1 "Multichannel DWDM Applications with Single-Channel Optical Interfaces," including the grid that defines channels and channel spacing. They also include a channel, optical supervisory channel (OSC), reserved for supervision and management purposes.

A WDM system consists of different functions that can be grouped in different devices:

- Transponder: It converts electrical to optical and vice versa.

- Multiplexer: It manages the combination of different signals, each with a wavelength, in the same pair of optical fibers (Figure 5.5).

- Demultiplexer: It separates wavelengths, sending them to the corresponding transponder.

- Repeater: It regenerates the signal to reach longer distances.

- Add-drop multiplexer: It integrates the multiplexer, demultiplexer, and transponder in the same device:
  - *Reconfigurable optical add-drop multiplexer (ROADM):* It is remotely configurable.

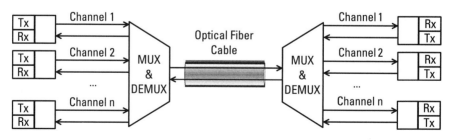

Figure 5.5   WDM.

• *Optical cross-connect (OXC):* Including some basic routing capabilities.

5.1.4 Optical Transport Network (OTN)

OTN is "composed of a set of optical network elements connected by optical fibre links, able to provide functionality of transport, multiplexing, routing, management, supervision and survivability of optical channels carrying client signals...." OTN is mainly described in ITU-T Recommendations G.872 "Architecture for the Optical Transport Network (OTN)," G.709 "Interfaces for the Optical Transport Network," and G.798 "Characteristics of Optical Transport Network Hierarchy Equipment Functional Blocks."

OTN is often referred to as a digital wrapper, allowing different services to be transparently transported. With this purpose, OTN defines a layered architecture with three levels (Figure 5.6):

- Digital layers:
    - Optical channel data unit (ODU). It provides services for end-to-end transport.
    - Optical channel transport unit (OTU). It provides transport services for ODU.
- Optical channel layer (OCh).
- Media layer.

Based on this architecture, OTN is flexible enough to transport application packets and also to encapsulate any kind of traffic (Figure 5.7) from other transport technologies such as WDM, SDH/SONET, and Ethernet. OTN provides fault management, performance monitoring, and protection mechanisms. Table 5.3 shows the list of available ODU services and their bit rates.

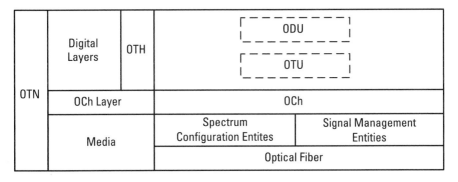

Figure 5.6 OTN-layered architecture.

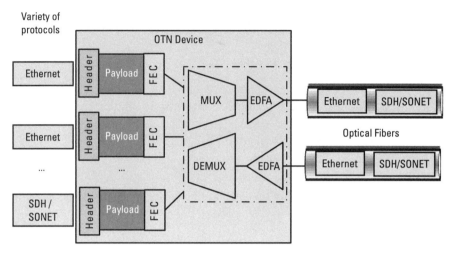

Figure 5.7   Transport technologies integration.

Table 5.3
OTN Interfaces

Service	Bit Rate (Gbps)
ODU0	1.25
ODU1	2.5
ODU2	10
ODU3	40
ODU4	100

The main advantages [4] of OTN over SDH/SONET are based on a stronger forward error correction (FEC), transparent transport of high data rate client signals, improved scalability, and tandem connection monitoring.

5.1.5   Radio Transmission

This section complements the vision on TDM, SDH/SONET, WDM, and OTN with a reference to radio used as media for transport.

When referring to radio transmission, the term radio link (defined in ITU-R V.573, as "a telecommunication facility of specified characteristics between two points provided by means of radio waves.") is used. When two or more links are connected, the term radio-relay system is used (defined in ITU-R V.573, as "Radio communication system between specified fixed points operating at frequencies above about 30 MHz which uses tropospheric propagation and which normally includes one or more intermediate stations."). Typical distances covered by individual radio links range from a few kilometers to as far as 100 km. If higher distances need to be covered, a chain of radio links (a radio-relay system with intermediate stations) needs to be installed.

Radio link-based transport does not have a standard definition of the *aggregate* interface: there is no standard that could be used in a radio link for the connection of the two ends on the radio interface, and any radio link terminal needs to communicate with its counterpart from the same vendor. Fortunately, tributaries (i.e., the interfaces for services) do usually follow TDM, SDH/SON-ET, and Ethernet standards for electrical interfaces. In terms of transmission capacity available, it can be generally stated that radio capacity is not as large as that of optical fiber transport.

Radio technologies used for transport can be broadly classified in two groups depending on the transmission context, namely terrestrial and satellite telecommunications.

From the frequency use perspective, radio technologies for transmission can be further divided in two groups: frequencies below 1 GHz [typically ultrahigh frequency (UHF)], and above 1 GHz (normally referred to as microwaves). Virtually all satellite transmissions are above 1 GHz.

There are three reasons to set the limit at 1 GHz. The first one is historical, as bands below 1 GHz were the first ones to be allocated by ITU-R for technology reasons (i.e., availability of associated electronics). The second one is related to the propagation characteristics, because above 1 GHz, the attenuation, due to atmospheric effects such as hydrometeors and others (obstacles, distances, and so forth), is higher. The third one is based on the characteristics of available bands, which are wider above 1 GHz, and consequently better prepared to accommodate higher data rate transmission.

Depending on the frequency and distance, data rates from 100 Mbps up to higher than 1 Gbps can be achieved with radio links. The lower the frequency, the higher the distance and the lower the capacity; the higher the frequency, the higher the capacity but the lower the distance.

Terrestrial radio links are common for data transport inside a certain region or country. Global transport is achieved with satellite telecommunications. Satellite can be used where other technologies are not present (e.g., remote locations such as solar or wind farms, hydro storage). When used for transport purposes, the geostationary orbit is the most common, as it allows a radio-relay system with fixed antennas.

Satellite networks use satellites as radio repeaters with a wider visibility and coverage than the terrestrial ones. However, satellites require huge investments, have a limited lifespan and the ground terminals are typically more expensive than their terrestrial counterparts. From the performance point of view, latency is high.

5.2   Switching and Routing Technologies

Switching and routing technologies are used to drive traffic, based on devices' addresses, from the origin to the destination on top of transport technologies or physical media within a telecommunication network. Each packet is directed through the correct link or hop to reach the next node of the network. The process is iterated until the final destination device (address) is reached. Originally designed for use in local area networks (LANs), with the evolution of the Internet and the technologies themselves, they have evolved to be also used in wide area networks (WANs).

5.2.1   Ethernet

Ethernet, as defined in IEEE 802.3-2012, "is an international standard for Local and Metropolitan Area Networks (LANs and MANs), employing CSMA/CD as the shared media access method and defining a frame format for data communication, intended to encompass several media types and techniques for a variety of MAC data rates."

Ethernet covers the two lowest layers of the OSI model, splitting the data link layer (DLL) into two sublayers, the MAC layer (included in Ethernet specification) and the logical link control (LLC) layer, which is defined in ISO/IEC 8802-2:1998 "Information technology - Telecommunications and Information Exchange Between Systems - Local and Metropolitan Area Networks - Specific Requirements - Part 2: Logical Link Control" (originally IEEE 802.2, withdrawn in 2010). Ethernet frame format is also defined in IEEE 802.3 (see Figure 5.8).

Ethernet defines multiple data rates over several physical media, from the 2.94 Mbps of the Experimental Ethernet [5] to early implementations with a 10 Mbps data rate over different media, such as coaxial cable (e.g., 10Base5 and 10Base2), twisted pair (10Base-T), or fiber optics (10Base-F). Nowadays Fast Ethernet provides data rates of 100 Mbps over twisted pair (100Base-T) or optical fiber, multimode (100Base-FX), or single mode (100Base-SX, 100Base-LX

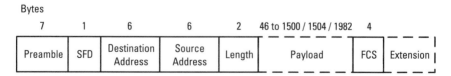

SFD–Start Frame Delimiter
FCS–Frame Check Sequence

Figure 5.8   IEEE 802.3 packet format.

and 100Base-CX). There are also interfaces for 1 Gbps, called Gigabit Ethernet, 10 Gbps (10 Gigabit Ethernet), 40 Gbps, and 100 Gbps.

The physical layer also defines an interface with the medium access control (MAC) sublayer that includes transmission and reception services, framing and contention resolution services, and a wait function for timing purposes.

The DLL defines two modes of operation, half-duplex and full-duplex. The half-duplex operation implements a CSMA/CD algorithm. The algorithm works as follows:

- A station waits for the medium to be silent before transmitting.

- The station sends the message and keeps listening to detect possible collisions.

- If the message collides with another message from other station, both stations intentionally transmit for an additional predefined period to ensure the propagation of the collision throughout the network.

- Both stations wait for a random amount of time (called back-off) before trying to transmit again the message.

The full-duplex operation does not need CSMA/CD implementation as it uses dedicated channels for transmission and reception respectively, allowing simultaneous communication between both stations.

The identification of the origin and destination of each Ethernet frame is achieved using Ethernet addresses based on the MAC addresses of the network interfaces and broadcasting protocols. When a new device is connected to an Ethernet network (also known as domain), there is an automatic discovery process and an adjustment of the network to identify the new entrant and in which interface it is available.

5.2.2   Internet Protocol (IP)

IP was originally designed to be used in the Internet, but its simplicity, scalability, and flexibility, together with the great success of the Internet, have turned IP into the most successful protocol at the network layer even for non-Internet applications.

As it is defined in RFC 791 (September 1981), "the internet protocol provides for transmitting blocks of data called datagrams from sources to destinations, where sources and destinations are hosts identified by fixed length addresses. The internet protocol also provides for fragmentation and reassembly of long datagrams, if necessary [...]". IP defines:

- An addressing method to uniquely identify a device (a network interface) and to facilitate the routing of datagrams;
- A packet format to be understood by any device in the network.

An IP address (IPv4) is a 32-bit binary number representing uniquely a network interface. These 32 bits are divided in four groups of 8 bits each and then converted to decimal notation. IP addresses are commonly represented (e.g., as 192.168.0.1).

IP protocol identifies origin and destination based on IP addresses instead of MAC addresses. Unlike Ethernet, broadcast protocols are not used to identify where these IP addresses are physically connected. Instead, the devices decide how to route an IP datagram based on the information they manage (routing tables), taking the decision on which is the next hop in the network for that destination. This information may be known by configuration (i.e., default routes) or dynamically learned through a routing protocol [e.g., routing information protocol (RIP), border gateway protocol (BGP), and open shortest path first (OSPF)].

The IP packet format shows two parts:

- Payload: It is the raw data of the transport layer protocol. The payload is transparently conveyed by the IP to the transport layer in the receiver.
- Header: It is the additional information added at the beginning of the payload by the IP in order to include the data needed for the packet to be routed through the network and understood by the destination host (Figure 5.9).

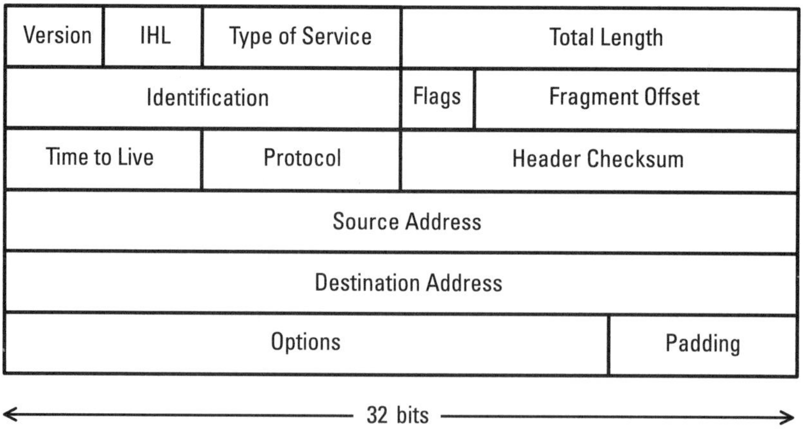

Version	IHL	Type of Service	Total Length	
Identification			Flags	Fragment Offset
Time to Live		Protocol	Header Checksum	
Source Address				
Destination Address				
Options				Padding

⟵——————— 32 bits ———————⟶

IHL–Internet Header Length

Figure 5.9   IPv4 packet header.

Due to the massive expansion of the Internet and the limited addressing space available in IPv4, the addressing capabilities of the protocol have been improved and a new version of IP, IPv6, exists. However, it is not yet widely spread (statistics in [6]).

5.2.3 Multiprotocol Label Switching (MPLS)

MPLS is defined by IETF in RFC 3031. The objective of this protocol is to simplify the mechanisms used by routers to forward packets to the next hop from the origin to the destination.

Packets in MPLS are assigned to a group by labeling them with a value that is used by the devices in the network to decide the path of the packet. Thus, the network layer header is not analyzed by any device in the path, as MPLS nodes only need to check the label to forward the packet to the corresponding next hop. Packet label can also be modified in this process if needed.

Each device in the MPLS network keeps a table of labels, storing the next device in the path for each particular label. A path defined between two endpoints is called the label-switched path (LSP) and there is a protocol called the Label Distribution Protocol (LDP) that is used to create and manage LSPs.

MPLS has advantages such as its processing simplicity and flexibility. MPLS requires a careful planning and configuration of labeling tables in the devices of the network. This situation gives the user more control of the network, leaving few decisions to be taken automatically by the devices. This ability to use traffic engineering is the main reason why MPLS is widely used as a packet switching solution, together with transport technologies such as WDM.

5.2.4 Transport-Adapted Technologies

Due to the evolution of electronics, together with the success of some switching and routing technologies mentioned in this section, some packet-based technologies have evolved to be used in the transport technologies domain. Two examples of this evolution are MPLS-TP (TP for Transport Profile) and Carrier Ethernet.

5.2.4.1 Multiprotocol Label Switching: Transport Profile (MPLS-TP)

MPLS-TP is a subset of MPLS protocols (among others), defined by IETF RFC 5654 and by ITU-T Recommendation G.8110.1 "Architecture of the Multi-Protocol Label Switching Transport Profile Layer Network." MPLS-TP is an evolution of MPLS that simplifies some features not needed in a transport technology and defines some new features to support traditional transport needs.

Among the new features in MPLS-TP [7], we have:

- Support for quality of service (QoS);

- Enhanced OAM functionality that allows monitoring and managing protection switching;

- Use of the generic associated channel (G-ACh) to support operations, administration, and maintenance (OAM) traffic (MPLS-TP separates data plane and control plane).

5.2.4.2 Carrier Ethernet

Carrier Ethernet was defined with the objective to develop an alternative to high-speed transport technologies, such as PDH or SDH/SONET. Although it was thought originally to cover metropolitan distances (in the order of tens of kilometers), it has evolved to be used for longer distances over optical fiber links.

Carrier Ethernet is defined and maintained by the Metro Ethernet Forum [8] and it is mainly an evolution of the Ethernet protocol providing connection-oriented services over an Ethernet network.

Reference [8] considered that the technology is able to offer QoS and scalable and reliable services that can be effectively managed. The advantages of this technology compared to other transport technologies are:

- Ethernet is part of the TCP/IP protocol stack, generally used for LANs, so it is simple to connect LANs to WANs.

- Devices and interfaces may be more cost-effective than alternative transport technologies.

- Data rate can be chosen granularly with the only limitation of the maximum speed defined in the protocol.

- The multiplexing of several connections over the same links is possible using virtual local area networks (VLANs).

5.3 Power Line Communication (PLC) Technologies

There are many different PLC technologies in the market. Some of them are better categorized as transport technologies and some others are categorized as local access technologies. PLC technologies are naturally well fitted to the electrical grid, although performance and availability are intrinsically influenced by events in it.

5.3.1 Broadband PLC Technologies

Starting in the second half of the 1990s, the telecommunications industry realized that PLC was a technology that could play a role in the diversification of technologies [e.g., radio, xDSL, hybrid fiber coaxial (HFC)] that helped the deregulation of telecommunications markets. Many utilities took active part in the emerging new telecommunication businesses, both as investors in telecommunication carriers (mobile operators) and also as promoters of their own infrastructure for telecommunications.

PLC was considered interesting because of the ubiquity of power grids, so a majority of customers could potentially be addressed with no additional cable deployments over an already deployed infrastructure. There were different initiatives to develop PLC technologies suitable for the access network; some of them were proprietary while others were supported by national or international open projects. Most of them reused concepts and technology already known and used in other systems [orthogonal frequency division multiplexing (OFDM), FDMA, TDMA, modulation, and coding schemes already developed and proven]; the interest was to achieve increasingly higher throughputs to compete with xDSL and HFC technologies that were being universally deployed. PLC finally became a success for the in-home domain where it is used to create a LAN at home through the electrical wiring [9, 10].

As a consequence, there are now several open and/or standard broadband (BB) PLC (BPL) technologies in the market that are summarized in the following sections as appropriate for the smart grids.

5.3.1.1 OPERA

Open PLC European Research Alliance (OPERA, [11]) was a project divided into two phases and funded by the EC's Sixth Framework Programme (FP6). It started in January 2004 with the objective "to offer low-cost broadband access service to all European citizens using the most ubiquitous infrastructure, which are power lines."

The purpose of the project was reformulated in its second phase (2007 to 2008) to find applications where PLC was especially useful, and to integrate with other complementary technologies such as radio. Among the different applications identified, smart grid was an obvious one due to its synergies with a technology that uses the power lines themselves. Thus, further to in-home applications, PLC was explored for telecontrol purposes in medium-voltage (MV) grid, remote metering (smart metering) in low voltage (LV) and smart grid backbone in the secondary substations (SSs) with BPL over MV cables.

OPERA produced prototypes and solutions ready to be delivered to the field. OPERA also produced an open specification in the context of a deployment model with three types of nodes, namely head ends (connecting the PLC network to other telecommunication technologies), repeaters (in the frequency or time domain), and customer premises equipment (CPE) to connect end-users.

The specification consists of the following layers:

- PHY layer: It is based on OFDM, using up to 1,536 subcarriers with configurable bandwidths of 10, 20, or 30 MHz and providing maximum data rates of 200 Mbps. Amplitude differential phase shift keying (ADPSK) is used with up to 1,024 points per constellation. Additional features of the physical layer are adaptive modulation, frequency notching (to avoid frequencies that could be prohibited due to local regulations), and FEC based on Reed-Solomon block codes and truncated four-dimensional trellis coded modulation.

- MAC layer: The MAC layer uses a TDMA mechanism, assigning, through a token, each device a time slot for transmission. The channel resources are managed by the head end device (acting as a master), guaranteeing QoS and providing a deterministic estimation on how much it will take a node to gain access to the channel.

- LLC layer: This layer performs three tasks: segment and assemble packets coming from the convergence layer, add a header (LLC delimiter) to each payload (forming a "burst"), and transmission control by assigning a sequence number to each burst and defining an acknowledgement mechanism.

- Convergence layer: This layer adapts Ethernet packets coming from external interfaces, making some modifications to its header for aspects such as priority, control, and others.

- Management layer: It defines a set of management protocols to control different features of the system, such as registration of new nodes, adaptive modulation, resource reservation, and collision clusters detection.

5.3.1.2 ITU-T G.hn

Following the evolution of the PLC industry, ITU-T started an effort to develop a recommendation for an in-home technology for different types of physical media (telephone lines, power lines, coax, and step-index polymer/plastic optical fibers), providing data rates of up to 1 Gbps. ITU-T established in 2006 the G.hn (Home Networking) Rapporteur Group, which issued ITU-T Recommendation G.9960 "Unified High-Speed Wireline-Based Home Networking Transceivers - System Architecture and Physical Layer Specification" in October 2009 and ITU-T Recommendation G.9961 "Unified High-Speed Wire-Line Based Home Networking Transceivers - Data Link Layer Specification" in June

2010 (the most recent versions are dated July 2015). Reference [12] addressed how to use G.hn for some smart grid applications [advanced metering infrastructure (AMI), in-home, and electric vehicles]. ITU-T Recommendations G.9962, G.9963 and G.9964 complete the picture. G.hn is supported by the HomeGrid Forum [10].

The specific characteristics of the PLC option in G.hn are:

- It uses OFDM with adaptive modulations from binary phase shift keying (BPSK) up to 4096-QAM and a frequency spacing between subcarriers of 24.4 kHz. A systematic quasi-cyclic low-density parity check block code (QC-LDPC-BC) encoder and a puncturing mechanism with various code rate options are used.

- It divides the network in domains with up to 16 different domains, typically (but not necessarily) in different physical media. Each domain has a designated domain master and up to 250 nodes, which are coordinated by the master. Some devices act as interdomain bridges to communicate devices in different domains and there is a global master to manage the resources for communications between domains.

- It defines a synchronized access to the media where transmissions are coordinated by the domain master and synchronized with the mains cycle to better manage grid-related noises.

- It defines two profiles to classify the nodes according to the functionality they implement. The standard profile is implemented by devices using two different band-plans, 50 MHz (1.8–50 MHz) and 100 MHz (1.8–100 MHz). The low-complexity profile (LCP) works in the 1.8–25 MHz range and only allows for ½ code rate and BPSK or quadrature phase shift keying (QPSK) modulations.

5.3.1.3 IEEE 1901

Parallel to the definition of G.hn standard by ITU-T, IEEE engaged with the definition of a BPL standard to be used both for in-home applications and for last-mile access connections. The IEEE 1901 standard was published in September 2010, defining both a physical layer (two different physical layers are included) and a MAC layer.

As for the physical layer, both layers included in the standard use frequencies between 2 MHz and 30 MHz, with an optional extension to 50 MHz, providing potential PHY data rates of up to 420 Mbps [13]. One of the PHY layers uses conventional OFDM and the other one employs wavelet OFDM. The first one is based on the definition made by the HomePlug Alliance [9], and the second one on the definition made by the HD-PLC Alliance [14].

The standard defines a common MAC layer that can manage both physical layers supported through intermediate layers, called physical layer convergence protocols. The access to the medium is accomplished by a TDMA mechanism based on CSMA/CA. In IEEE 1901, the building blocks of the network are referred to as the basic service set (BSS), also known as cell, and the device that manages this network is called BSS manager, which coordinates the devices in the network, referred to as stations.

5.3.1.4   Korea ISO/IEC 12139-1

Simultaneously with the other initiatives mentioned before, Korean Agency for Technology and Standards prepared a BPL standard, called initially KS X 4600-1, which was adopted by ISO and IEC in 2009 with the name ISO/IEC 12139-1. It defines both PHY and MAC layers to support both IEEE 802.3 and serial protocols as upper layers, in frequencies below 30 MHz.

As for the physical layer, it defines OFDM in frequencies between 2.15 MHz and 23.15 MHz with subcarrier spacing of 97.65625 kHz and adaptive modulation of differential BPSK (DBPSK), differential QPSK (DQPSK), or differential orthogonal PSK (D8PSK). A FEC mechanism is also defined based on Reed-Solomon and convolutional coding.

For the MAC layer, it defines a CSMA/CA mechanism to access the media. For addressing, it identifies each logical network or cell with a group identifier and each station of this group with a station identifier mapped to the corresponding MAC address, normally coming from the upper layer (IEEE 802.3). Request to send (RTS) and clear to send (CTS) mechanisms can be used to prevent collisions caused by hidden stations (STAs). Cell bridge stations are defined to communicate different cells.

The ISO/IEC 12139-1 MAC layer also defines a procedure for the stations to perform channel estimation and select the most suitable subcarrier mapping for a particular link and a mechanism to reserve slots for transmission to manage priorities.

5.3.2   Narrowband PLC Technologies

From 2005 onwards, based on the need to deploy smart metering, narrowband (NB) PLC gained new attention. This led to a new set of requirements for state-of-the-art PLC technologies that allowed the remote management (i.e., reading and controlling) of smart meters. Several initiatives were developed to define PLC technologies suitable for this purpose.

5.3.2.1   PRIME

PRIME stands for powerline intelligent metering evolution, and it is a narrowband high data rate (HDR) PLC technology that defines a three layers' architecture (PHY, MAC, and convergence) for "complexity-effective, data transmission over electrical power lines that could be part of a Smart Grid system." PRIME is defined and maintained by the PRIME Alliance [15], and it has been adopted by the ITU-T (ITU-T Recommendation G.9904) and included in the IEEE (IEEE 1901.2).

As for the PHY layer, PRIME defines a multicarrier OFDM strategy with 97 carriers in each of the eight independent channels between 42 kHz and 472 kHz. Channels can be used separately or grouped [16] to increase throughput.

PRIME also defines several modulation schemes, which are adaptively selected by network devices, in order to obtain higher bit rates or provide increased robustness as needed. The mechanism to perform this functionality is called PHY robustness management (PRM).

The PHY frame structures defined, type A and type B, are shown in Figure 5.10.

With this definition, a PRIME subnetwork, depending on the modulation scheme selected and the number of channels used, provides a raw data rate between 5.4 kbps and 1,028.8 kbps.

As for the MAC layer, a PRIME subnetwork is defined with two types of nodes, namely, base node (BN) and service node (SN). The BN is the unique master of the subnetwork, in charge of managing the network resources assigning them to the SNs registered in the subnetwork. SNs, once registered in a subnetwork, are managed by the BN. SNs can perform as terminals or switches. As terminals, they just provide communications for their application layer; as switches, they switch (repeat) other nodes' packets helping the BN to extend the connectivity of the subnetwork.

In terms of addressing, each PRIME node has a 48-bit MAC address, but in order to reduce overhead, the MAC address is not always transmitted in the packet header, only during the registration process.

PRIME offers an automatic repeat request (ARQ) mechanism. It is an end-to-end property of the connection, implemented through a 6-bit packet identifier, associated with each packet.

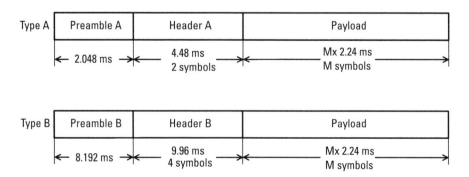

Figure 5.10   PRIME PHY frame structures (type A and type B).

As for the convergence layer, PRIME specifies three different layers (IPv4, IPv6, and IEC 61334-4-32) to support any smart grid application network layer.

An additional feature provided by the convergence layer is the segmentation and reassembly (SAR) mechanism, which is designed to control the size of the packet delivered to the MAC layer in order to avoid long packet data units (PDUs) and reduce the probability of collisions and impulsive noise interference.

A management plane is also available to operate the network.

5.3.2.2   G3-PLC

G3-PLC is a narrowband PLC technology that defines a three-layer architecture (PHY, MAC, and adaptation layer). It is defined and maintained by the G3-PLC Alliance [17], and it has been adopted by the ITU-T (ITU-T Recommendation G.9903) and included in the IEEE (IEEE 1901.2).

For the PHY layer, G3-PLC defines a multicarrier OFDM strategy with a set of configurable band plans in frequencies between 36 kHz and 487.5 kHz.

G3-PLC defines several modulation schemes, which are adaptively selected by network devices. The specification defines a mandatory rate ½ convolutional code. Data is additionally encoded with a shortened Reed-Solomon code. G3-PLC networks provide a raw data rate between 4.5 kbps and 298.2 kbps.

The MAC layer is based on IEEE 802.15.4. The adaptation layer is based on 6LoWPAN, which is defined by the IETF in RFC 6282. The 6LoWPAN is a standard designed for low rate wireless networks, and specifically featured to work over IEEE 802.15.4 MAC layer. With this objective, it defines encapsulation and header compression mechanisms to allow IPv6 packets to be exchanged over low data rate networks.

5.3.2.3   IEEE 1901.2

Following the path started by both PRIME Alliance and G3-PLC Alliance, together with the effort to standardize a BB PLC technology, IEEE started an action to develop a standard for low-frequency (LF) (less than 500 kHz) NB PLC for smart grid applications. IEEE 1901.2 includes both PRIME and G3-PLC as 1901.2-compliant.

The PHY layer is OFDM-based (from 10 to 490 kHz, and different band plans), with different modulation schemes (DBPSK, DQPSK, D8PSK, and, optionally, coherent BPSK, QPSK, 8PSK, and 16QAM). MAC layer leverages IEEE 802.15.4 services and primitives. The architectural reference model can be based either on L2 mesh or L3 routing [e.g., Routing Protocol for low power and lossy networks (RPL)].

5.3.2.4 G.hnem

Similar to the IEEE, the ITU-T defined an evolution of G.hn specifically designed to address the need of smart grids in general and smart metering in particular. The evolution is called G.hnem, where "em" stands for energy management. The set of standards includes ITU-T Recommendation G.9902, together with G3-PLC (ITU-T Recommendation G.9903) and PRIME (ITU-T Recommendation G.9904) covering MAC layers and above. ITU-T Recommendation G.9901 covers the PSD and PHY specific aspects in the different frequency plans.

The network architecture is similar to that of G.hn. It is divided in logical domains with unique identifications in the network. A particular single node can belong to only one domain. The nodes in different domains communicate with each other via interdomain bridges. Alien ITU-T G.9902 domains and other G.9902 networks may be communicated via L3 bridges. A global master coordinates the different domains in the same network.

The PHY layer is OFDM-based with different band plans (including FCC band up to 490 kHz), with different modulation schemes. The MAC layer is based on CSMA/CA, and L2 or L3 mechanisms are supported.

5.4   Radio Systems

Radio propagation characteristics in the different frequency bands allow the existence of a variety of radio systems with different applications and scope, some of them specifically suitable for smart grids.

Radio systems have different origins, which explain their capabilities. PMR and cellular radio systems come from traditional radio applications for voice purposes and have evolved to provide data transport capabilities. PMR is usually associated to private use, and cellular to public systems. Wireless systems compliant to IEEE standards stem from the adaptation of wireline packet-based technologies (i.e., Ethernet) to wireless transmission; these are usually standards for private use and the difference among them resides in their expected range: 802.16 covers distances of kilometers, 802.11 covers distances of hundreds of meters, and 802.15 covers distances of tens of meters.

5.4.1   Private Moblie Radio (PMR)

PMR (sometimes also referred to as Professional Mobile Radio [18]) refers to a set of radio systems normally providing low data rates and dedicated to private use. These include applications for utilities, public protection and disaster relief (PPDR), and car fleet communications. PMR systems were the precursors of current public systems: they started with simple amplitude modulation (AM), adopting frequency modulation (FM) for better voice quality; eventually they evolved from analog to digital, including advanced capabilities such as trunking to increase the efficiency of the system. In some cases, this evolution enabled the standardization of new systems.

PMR networks consist of a number of base stations and a larger number of terminals (fixed or mobile) connected to them. PMR systems generally provide services to closed groups of users, offering group calls with push-to-talk; the system is designed to offer short call setup in channels allocated for the duration of short calls, and sometimes offer the direct mode operation in which terminals can communicate among them without the help of the base stations. Many of these systems work in bands such as 80, 160, and 450 MHz to have a long range, with channels from 6.25, 12.5, and 25 kHz [19].

A list of the best known PMR systems follows:

- Land Mobile Radio (LMR): It is a generic reference to a set of systems used for wireless communications between devices, normally used by vehicle fleets or emergency organizations. In the context of the ITU, LMR is mentioned in several recommendations.

- Terrestrial trunked radio (TETRA): It is a standard from ETSI [20]. It provides a scalable architecture that allows local and nation-wide coverages. TETRA has evolved with Release 2, which includes the TETRA enhanced data service (TEDS) trying to guarantee the evolution towards higher speed applications.

- Digital mobile radio (DMR): It is a standard from ETSI aiming to include the technology progress and the bandwidth efficiency in the transition from analog to digital in the typical voice applications (being a digital system, nothing prevents its use for data transmission purposes). An introduction to the system can be found in ETSI TR 102 398.

- Project 25 (P25): It is an American initiative producing a set of standards within the Telecommunications Industry Association (TIA) to enable the interoperability among multiple manufacturers in the air interface between P25 radios and infrastructure. These standards are mostly developed within the TIA's Engineering Committee TR-8. TIA-102 series standards are within the Association of Public Saftey Communications Officials (APCO) Project 25.

5.4.2 WiMAX

WiMAX is a wireless technology specified as IEEE 802.16 for broadband wireless metropolitan area networks (WMANs).

IEEE 802.16 has evolved over time, consolidating several amendments, as it is usual in IEEE procedures. Table 5.4 shows a list of active standards of IEEE 802.16 WG.

IEEE 802.16-2012 specifies the air interface, PHY layer, and MAC layer, both for fixed access and mobile point-to-multipoint broadband wireless access systems. The MAC supports the WirelessMAN-SC (SC, Single Channel), WirelessMAN-OFDM, and WirelessMAN-OFDMA PHY specifications, each suited to a particular operational environment.

The PHY layer defines two different ranges of frequencies: between 10 and 66 GHz, requiring line of sight (LOS), and below 11 GHz, with a not-so-strict requirement of LOS. The lower frequency range, due to its propagation characteristics, is the one with more market penetration for access networks where near LOS may be feasible; license-exempt bands are also possible. For this sub-11 GHz range, the standard defines OFDM and OFDMA (OFDM Access; to assign sets of subcarriers to different end-users) schemes with selectable modulations between QPSK, 16-QAM, and 64-QAM.

The IEEE 802.16 connection-oriented MAC layer defines an architecture where the base station is the manager of the network. For the downlink, the base station transmits on a point-to-multipoint basis, while for the uplink the users are assigned time slots and subcarriers using a contention period for negotiations depending on their requirements of QoS. The MAC layer also al-

Table 5.4
IEEE 802.16 Standards

Standard	Title
802.16-2012	Air Interface for Broadband Wireless Access Systems
802.16p-2012	Amendment 1: Enhancements to Support Machine-to-Machine Applications
802.16n-2013	Amendment 2: Higher Reliability Networks
802.16q-2015	Amendment 3: Multi-tier Networks
802.16.1-2012	WirelessMAN-Advanced Air Interface for Broadband Wireless Access Systems
802.16.1b-2012	Amendment 1: Enhancements to Support Machine-to-Machine Applications
802.16.1a-2013	Amendment 2: Higher Reliability Networks
802.16.2-2004	Recommended Practice for Local and Metropolitan Area Networks - Coexistence of Fixed Broadband Wireless Access Systems
802.16/ Conformance04-2006	Standard for Conformance to IEEE 802.16 - Part 4: Protocol Implementation Conformance Statement (PICS) Proforma for Frequencies below 11 GHz
802.16k-2007	Standard for Local and Metropolitan Area Networks: Media Access Control (MAC) Bridges - Amendment 2: Bridging of IEEE 802.16

Source: [21].

lows for multihop operation, allowing the use of one or more relay stations to get access to distant stations.

5.4.3   ZigBee

ZigBee refers to a system for various applications needing inexpensive and power-efficient devices within the context of the WirelessPANs, that is, networks covering short distances (hundreds of meters) and low data-rates (kbps). ZigBee uses IEEE 802.15.4 in the PHY and MAC layers and covers all the layers up to the application layer. ZigBee Alliance [22] is the industry consortium managing the definition of the system and supporting the different applications such as smart homes and appliances and smart metering.

From a system-level perspective, IEEE 802.15.4 networks are intended to need little or no infrastructure. Two different entities are defined; one acts as the PAN coordinator and the other has a reduced set of functionalities to need a minimum set of resources. Any IEEE 802.15.4 needs a PAN coordinator and its topology can be built upon a star or a peer-to-peer structure. Star networks are created by PAN coordinators and are independent among them; peer-to-peer networks may create cluster trees with different coordinators, where each device can communicate with any other device within its range (mesh capabilities).

The PHY layer in IEEE 802.15.4 covers different license-exempt bands worldwide including 868–868.6, 902–928 MHz and 2,400–2,483.5 MHz, for a raw data-rate of up to 250 kbps. The standard covers different PHYs adopted over time and this is why a set of different modulations can be found in the standard, being spread spectrum the most popular option. The frequency and rest of the channel conditions (e.g., channel limits and transmit power) are affected by national regulations worldwide. The standard specifies some performance conditions to be met (e.g., receiver sensitivity better than −85 dBm for a certain packet error rate conditions).

The MAC layer in IEEE 802.15.4 specifies the use of either a CSMA/CA or an ALOHA mechanism to share the medium. ALOHA is just appropriate for low-load networks. CSMA/CA is used to access the medium trying to avoid collisions. A super-frame is defined within two beacon transmissions, and with an active period (16 time slots) and an optional inactive period; the active period is used combining a contention access and a contention free period. The contention-free period can be used if QoS need to be met. The inactive period allows the coordinator to enter in low-consumption mode.

Outside the scope of IEEE 802.15.4, ZigBee defines application standards. One of these is Smart Energy, where different versions (profiles) evolve to suit the needs of the industry. Smart energy profile (SEP) 1.x is evolving into SEP 2.0 to use ZigBee IP stack through the 6LoWPAN protocol encapsulating ZigBee packets on IPv6 datagrams.

IEEE 802.15.4g is an amendment to IEEE 802.15.4 with a self-declared PHY specifically defined for Smart Metering utility networks. Wi-SUN alliance [23] uses this standard. Multirate and multiregional FSK, OFDM (MR-OFDM), and offset QPSK (MR-O-QPSK) are defined in bands ranging from 169 MHz to 2,450 MHz, together with the necessary MAC adaptation.

5.4.4 Wi-Fi

Wi-Fi refers to a set of technologies covered by the IEEE 802.11 standards for WLANs and promoted by the Wi-Fi Alliance [24]. In 1990, IEEE 802.11 initiated the concept of WLAN, and in 1997 the first version was published. Now a single standard, IEEE 802.11-2012 includes all previous well-known amendments (IEEE 802.11a, b, and g).

IEEE 802.11 addresses wireless connectivity for fixed, portable, and moving stations within a local area. The typical use-case of Wi-Fi is the connectivity for PCs at the office or at domestic environments. The distribution system architectural concept can be used to interconnect infrastructure (basic service sets) and create large coverage networks called extended service sets. IEEE 802.11 can also support the creation of mesh radio topologies. IEEE 802.11 networks have eventually evolved including QoS and authentication capabilities.

The protocol stack defined by IEEE 802.11 includes the PHY and the MAC layer, with a physical layer convergence procedure between both of them to make the MAC layer independent from the multiple physical layers supported by the protocol:

- Direct spectrum spread spectrum (DSSS) in the 2.4-GHz band DBPSK and DQPSK to provide the 1 Mbps and 2 Mbps data rates;
- OFDM in the 5 GHz band with communication capabilities of 6, 9, 12, 18, 24, 36, 48, and 54 Mbps (using BPSK or QPSK or using 16-QAM or 64-QAM, and convolutional coding FEC);
- Frequency hopping spread spectrum (FHSS) in the 2.4 GHz band (obsolete);
- High-rate DSSS (HRDSSS);
- Extended-rate PHY;
- Infrared.

CSMA/CA is used at the MAC layer. A distributed coordination function exists, on top of which, both a hybrid coordination function for advanced mechanisms of contention to allow network QoS and a mesh coordination function to be used in this kind of networks are defined.

5.4.5   Cellular

During the last two decades, cellular systems have experienced a huge evolution to support ubiquitous voice and data connectivity. Last generation technologies have managed to support the deployment of systems offering tens of megabits per second per user with latencies in the order of high tens of milliseconds. The main technical improvement of modern cellular systems technology derives from their ability to increase the spectral efficiency (i.e., the amount of data effectively exchanged per unit of bandwidth).

The technology has enabled the deployment of different generations (broadly categorized as 2G/3G/4G) of coexisting systems that are commercially available. These systems coexist within the same telecommunication service provider (TSP) public system and are used transparently to the end-user.

Among the variety of cellular systems and standards, the most relevant will be mentioned. These technology options are present worldwide, but with different penetrations. From 3G onwards, they usually fall within the scope of the international moblie telecommunications (IMT) of the ITU-R, both IMT-2000 and IMT-Advanced.

IMT-2000 comprises 3G proposals from global partnership projects and recognized standards development organizations (SDOs) with radio physical layers based on different medium access and modulation alternatives, as covered in ITU-R Recommendation M.1457. The different options are CDMA direct spread, CDMA multicarrier, CDMA TDD, TDMA single-carrier, FDMA/TDMA, and OFDMA TDD WMAN.

IMT-Advanced recognizes two systems in the fourth generation (4G) domain. The systems are LTE-Advanced (developed by 3GPP as LTE Release 10 and Beyond) and WirelessMAN-Advanced (developed by IEEE as the WirelessMAN-Advanced specification incorporated in IEEE 802.16.1).

Vendors and standardization associations worldwide have deeply conditioned the availability of each group of systems in the different countries. The Third Generation Partnership Project (3GPP) [25] and Third Generation Partnership Project 2 (3GPP2) [26] are two of the relevant organizations with interest in these standards. Both organizations were created almost in parallel to ensure the evolution of the existing second generation (2G) systems, namely, European GSM-based systems and North American/Asian ones.

To cover all the standards worldwide is out of the scope of this chapter. The evolution of the European-origin standards is a representative example:

- Global system for mobile (GSM) communications: It was standardized by ETSI in 1990 to provide quality mobile voice services within scalable cellular networks. Data transmission services supported were basic and offered symmetric 9.6 kbps.

- General packet radio service (GPRS): It was the evolution of GSM, approved by the ETSI in 1998 and intended to provide asymmetric data services, without great changes in the radio access interface. The maximum allowable data rate in a GPRS duplex carrier is around 170 kbps, but typical commercial services support averages of 10 kbps. Together with enhanced data rates for GSM evolution (EDGE), which offered improved data rates once the radio interface modulation was modified, it is considered to belong to 2.5G.

- Universal mobile telecommunications system (UMTS): It is a 3G system defined by 3GPP within the IMT-2000 initiative, using in its first releases wideband code division multiple access (WCDMA) and including several air interfaces. UMTS is not a unique system, as it follows different parallel releases, from the initial Release 99, to the subsequent Release 4 to Release 14. UMTS evolved to include different features to increase packet data performance in a set known as high-speed packet access (HSPA); first high-speed downlink packet access (HSDPA) came in Release 5, and high-speed uplink packet access (HSUPA) came in Release 6. The combination of both is known as HSPA. HSPA evolution (also known as HSPA+ and evolved HSPA) came in Release 7 with further improvements in later releases.

- Long-Term Evolution (LTE): UMTS evolved releases gave birth in 2008 to the Release 8, known as LTE. LTE is a 4G system specified over an OFDMA radio access scheme to allow for theoretical peak data rate in the downlink higher than 100 Mbps. LTE is not just a new radio access but a set of new elements within an evolved packet system. The evolution of LTE continues under the name of LTE-Advanced from Release 10 onwards.

5.5 Wireline Systems

Due to the increasing demand of residential Internet access over the last two decades, TSPs have deployed technologies to provide access to the Internet with considerable bandwidth to areas where potential demand could exist. These solutions, many of them based on wireline technologies (over existing or even new laid cables), have evolved to cost-effectively provide data rates of up to 100 Mbps. This section covers best known wireline technologies, namely xDSL (over telephone lines), HFC (over optical fiber and coaxial cables), and FTTx (over optical fiber cables).

5.5.1 DSL

DSL refers to technologies designed to transport data over existing twisted telephone pairs, sharing the medium with Plain Old Telephone Service (POTS). Coexistence with the telephone service is possible due to the frequency division multiplexing nature of DSL systems, as telephone service uses typically frequencies between 300 and 3,400 Hz (voice band is considered below 4 kHz), while data transport uses different frequencies that can be filtered apart. DSL system use frequencies up to 30 MHz. The work on DSL has been consolidated under the umbrella of ITU-T SG15\Q4, where ETSI and ANSI have strongly contributed.

There is a variety of different systems among DSL technologies (this explains the use of the xDSL acronym). According to ITU-T Recommendation G.Sup50, "Overview of Digital Subscriber Line Recommendations," ADSL, HDSL, SHDSL, and VDSL are the main technologies in the xDSL domain. Table 5.5 shows the list of ITU-T recommendations. ADSL offers asymmetric traffic data rate (much higher in download) and other technologies such as VDSL may offer symmetric traffic transport. HDSL and SHDSL have been used to connect network elements over copper pairs; other alternatives such as ITU-T Recommendations G.998.1, G.998.2, and G.998.3 exist for this same purpose.

ADSL and VDSL are the most used technologies by TSPs to provide Internet access to their customers; ADSL from local exchanges (see Figure 5.11) and VDSL combined with optical fiber deployments to get closer to customers. ADSL is standardized by ITU-T Recommendation G.992 series (1 to 5) and in its first definition supported a minimum of 6.144 Mbps downstream and 640 kbps upstream while in the last version, it supports a net data rate ranging up

Table 5.5
ITU-T xDSL Recommendations

xDSL	ITU-T Recommendation	Description
HDSL	G.991.1	High bit rate digital subscriber line (HDSL) transceivers
SHDSL	G.991.2	Single-pair high-speed digital subscriber line (SHDSL) transceivers
ADSL	G.992.1	Asymmetric digital subscriber line (ADSL) transceivers
ADSL	G.992.2	Splitterless asymmetric digital subscriber line (ADSL) transceivers
ADSL2	G.992.3	Asymmetric digital subscriber line transceivers 2 (ADSL2)
ADSL2	G.992.4	Splitterless asymmetric digital subscriber line transceivers 2 (splitterless ADSL2)
ADSL2+	G.992.5	Asymmetric digital subscriber line transceivers 2 (ADSL2) – Extended bandwidth (ADSL2plus)
VDSL	G.993.1	Very high-speed digital subscriber line (VDSL) transceivers
VDSL2	G.993.2	Very high-speed digital subscriber line transceivers 2 (VDSL2)

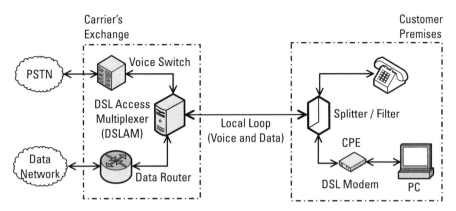

Figure 5.11 ADSL service typical access.

to a minimum of 16 Mbps downstream and 800 kbps upstream. VDSL is standardized by ITU-T Recommendation G.993 series (1 and 2), including two evolutions of the technology, which is now in its second version, called VDSL2 (G.993.2), able to provide 100 Mbps both in uplink and in downlink thanks to the use of up to 30 MHz of the spectrum.

From the technical point of view, all DSL technologies define a physical layer based on twisted pairs dedicated for each customer, establishing a link between the DSLAM (within the public carrier premises' local exchange) and the CPE (where the service is delivered). DSL technologies use a DMT multiplexing technique, a multicarrier one conceptually similar to OFDM. Each tone is modulated in QAM, and the technology allows flexible assignment of frequencies. In addition, FEC techniques are used to maximize robustness of the transmissions.

The differences between the versions of the technology are mainly the frequency and the number of carriers used (to increase data rate), as well as the mechanisms defined to allow this use of higher frequencies reducing the signal-to-noise ratio (SNR) needed to decode each frame in the receiver. These mechanisms are limited by the cable infrastructure, both the diameters of the cables and the number of twisted pairs packed together, due to the cross-talk among them. These constraints limit to lower distance ranges as the data rate increases (the maximum distance considered in ADSL is around 5 km).

Apart from the physical layer, all DSL technologies define a protocol stack composed by three sublayers situated between the physical layer and the transport protocol on top of the stack (some recommendations follow this more loosely). First of all, a physical media-dependent sublayer is defined and, over it, a physical media-specific transmission convergence function. Over these two sublayers, a Transport Protocol-specific transmission convergence function is defined for data transmission. Operation and management is achievable through a management protocol-specific transmission convergence function, with access to some control features of the different layers (e.g., power management).

5.5.2 HFC

HFC defines an access technology based on the combination of optical fiber and coaxial cable networks. Its origin is based on the evolution of the utilization of existing coaxial cables deployed for distributing cable TV (CATV and CCTV). Coaxial cables are used closer to the end-customers, and the media may support bidirectional voice and data services as well as the original TV broadcasting services. The same cable is shared by several customers reducing the maximum individual capacity available, and this is why TSPs have deployed fiber closer to the end-customer to reduce the number of them sharing the medium. The typical architecture in the coaxial cable domain has its origin in the cable modem termination system (CMTS) and ends in the cable modem in customer premises, as shown in Figure 5.12.

The dominant technology used over coaxial cables is called Data Over Cable Service Interface Specification (DOCSIS). CableLabs [27] is the reference consortium of the technology over coaxial cables offering a technology forum and certification programs to achieve interoperability; the Society of Cable Telecommunications Engineers (SCTE) and other entities worldwide (e.g., ETSI) do also have an active role as there are different regional particularities in these networks. DOCSIS has been standardized by ITU-T and it has evolved in different generations from DOCSIS 1.0 to 3.1 (Table 5.6), while DOCSIS 3.1 cannot be explicitly found in ITU-T yet.

The recommendations cover different band plans and particularities in the different world regions. Specifically, there are three different parts covering the work worldwide and the different band-plans (Europe, North America, and Japan); ITU-T J.122 includes three options for physical layer technology with different channel spacing for download multiprogram television distribution (channels of 8 MHz in Europe; 6 MHz in North America and Japan), and upstream transmission in different band ranges depending on the region and

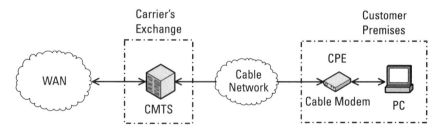

Figure 5.12 HFC data network access.

Table 5.6

DOCSIS ITU-T Recommendations

DOCSIS	ITU-T Recommendation
1.0	J.112 Annex B (superseded version)
1.1	J.112 Annex B
2.0	J.122
3.0	J.222.0, J.222.1, J.222.2, and J.222.3; J.201
3.1	Not yet as an ITU-T Recommendation

the version of the standard (see details in the ITU recommendations). ITU-T J.83 covers the definition of the framing structure, channel coding and modulation for digital multiprogram signals for television (including sound and data services) distributed by cable networks.

From the technical standpoint, DOCSIS shows a frequency division organization, with an upstream band and a downstream band. Of late, the interest for data in cable networks has grown, and the different standard versions show this effort to increasingly produce systems with a higher bandwidth capacity (from the 5 Mbps upstream in DOCSIS 1.0 to 30 Mbps in DOCSIS 2.0, that grow virtually "limitless" in DOCSIS 3.1; however, the objective is to get closer to fiber-based access offering 100 Mbps). Upstream band goes up to 85 MHz in DOCSIS 3.0 for some of the band plans while this limit may reach 204 MHz in DOCSIS 3.1. Access to the medium can be gained with a combination of FDMA/TDMA and FDMA/TDMA/S-CDMA. Modulations range from QPSK to 128 QAM and above. DOCSIS standards have introduced improvements in QoS, resource sharing, and frequency domain, progressively improving performance. DOCSIS systems limit latency and thus the extent where it can be used. The final performance depends on the number of users (resource sharing), and the noises and interferences due to the users in the same cable.

DOCSIS is designed to be simple and transparent. At network layer, DOCSIS requires the use of IPv4 or IPv6 for data transport between the CMTS and the CM. CMTS can perform MAC bridging or network layer routing, while the cable modem only performs MAC layer bridging of data traffic. Both CMTS and cable modem are network and transport layer aware (QoS and packet-filtering purposes). Additionally, DOCSIS requires use of the following higher-layer protocols [simple network management protocol (SNMP), trivial file transfer protocol (TFTP), dynamic host configuration protocol (DHCP)] for operation and management of the devices.

5.5.3 FTTx

FTTx is the acronym for fiber to the x, with x being any of the premises between the network core and the end-customer (see Table 5.7).

Table 5.7

FTTx Deployment Types

FTTx	Meaning
C	Curb or Cabinet
O	Office
B	Building
N	Node
P	Premises
H	Home

FTTx term comprises any telecommunication architecture in which the optical fiber is deployed to bring the optical network of the TSPs as close as possible to the end-user. Reference [28] referred mainly to FTTH, FTTB, FTTC, and FTTDp (Table 5.7) (Dp, Distribution point) and considers the first two as the target of the technology; however, the rest may exist as intermediate situations like those combining DSL and coaxial cable transmission techniques. FTTx deployments typically refer to the passive use of the technology, as opposed to active networks where electrical intermediate regeneration could be achieved with electronic elements (switches, routers, Ethernet in the first mile and point-to-point solutions typically). Passive networks, consisting of optical line terminations (OLTs) in the carrier's premises, optical network terminations (ONTs) for end-customers, and optical splitters, are deployed with a point-to-multipoint structure and network protocols run over them.

Optical networks do not only include network protocols but also the different optic elements in it, including design and maintenance. There are several organizations involved in the definition of standards related to the optical network infrastructure and protocols. ITU-T, through its SG 15, has published several standards within the G series; some of them refer to the optical fiber, such as G.65x standardizing types of fiber cables, and some other to the transmission technology itself with G.98x standardizing different versions of passive optical networks (PONs); L series deals with the physical infrastructure itself (construction, installation, and protection of cables and other elements of outside plant). Also CENELEC (TC86 and TC215), ETSI, IEC (TC86), ISO/IEC JTC 1/SC 25, IEEE [P802.3; IEEE 802.3ah, ethernet PON (EPON), and 802.3av (10G EPON)] are the main standard references. Broadband Forum [29] and others provide standards related to fiber optics and in particular to FTTx architectures. Apart from these organizations dedicated to telecommunications in general, it is also worth to mention the work of FTTH Councils to "support the rollout of fiber access networks to homes and businesses" [30].

ITU's broadband PON (BPON) (ITU-T G.983 series, 622 Mbps), Gigabit PON (GPON) (ITU-T G.984 series, 1 Gbps), 10 Gigabit PON (XG-PON) (ITU-T G.987 series, 10 Gbps) and next-generation passive optical network

(NG-PON2) (ITU-T G.989 series, 40 Gbps) standards, and IEEE's EPON (IEEE 802.3ah, 1 Gbps) and 10G-EPON (IEEE 802.3av, 10 Gbps) coexist in the different world regions. EPON and GPON have a common basis [31] in the general concepts of the ITU-T G.983 (operation, ODN framework, wavelength plan, and application), but significant differences. Thus, while EPON is based on IEEE 802.3 Ethernet, from where it was modified to support point-to-multipoint, GPON is fundamentally a transport protocol (relying on SDH/SONET and Generic Transport Protocol for Ethernet transport). With EPON, the traffic is broadcasted downstream and passively divided in the optical splitters; in upstream, the traffic has a TDMA scheduling. GPON is synchronous by nature, and the traffic is divided by the optical splitters in downstream, and scheduled with TDMA upstream. On the side of the common aspects, both standard streams need to uniquely identify the OLTs, support their automatic discovery, encrypt network traffic, and allocate TDM slot in upstream to each OLT (dynamic bandwidth allocation).

References

[1]   Horak, R., *Telecommunications and Data Communications Handbook*, New York: Wiley, 2007.

[2]   JDSU, *SDH Pocket Guide*, 2013. Accessed March 24, 2016. http://www.viavisolutions. com/sites/default/files/technical-library-files/sdh_pg_opt_tm_ae.pdf.

[3]   Coriant, *Coriant and Orange Set World Records for Optical Transmission Capacity and Distance Using State-of-the-Art Modulation Technology in Multi-Terabit Field Trial,* June 2015. Accessed March 24, 2016. https://www.coriant.com/company/press-releases/Coriant-and-Orange-Set-World.asp.

[4]   Walker, T., *OTN Tutorial,* Telecommunication Standardization Sector of International Telecommunication Union (ITU-T), 2005. Accessed March 24, 2016. https://www.itu. int/ITU-T/studygroups/com15/otn/OTNtutorial.pdf.

[5]   Spurgeon, C. E., and J. Zimmerman, *Ethernet: The Definitive Guide*, 2nd ed., Sebastopol, CA: O'Reilly, 2014.

[6]   Internet Society, "IPv6 Statistics," 2015. Accessed March 24, 2016. http://www.internet-society.org/deploy360/ipv6/statistics/.

[7]   *Understanding MPLS-TP and Its Benefits*, Cisco Systems, 2009. Accessed March 24, 2016. http://www.cisco.com/en/US/technologies/tk436/tk428/white_paper_c11-562013.pdf.

[8]   MEF, "About MEF," 2016. Accessed March 24, 2016. http://www.mef.net/about-us/mef-overview.

[9]   HomePlug Alliance, "About the HomePlug® Alliance," 2016. Accessed March 24, 2016. http://www.homeplug.org/alliance/alliance-overview/.

[10]   HomeGrid Forum, "About HomeGrid," 2014. Accessed March 24, 2016. http://www. homegridforum.org/content/pages.php?pg=about_overview.

[11] Open PLC European Research Alliance (OPERA) for New Generation PLC Integrated Network, "Project Details," 2008. Accessed March 21, 2016. http://cordis.europa.eu/project/rcn/71133_en.html. http://cordis.europa.eu/project/rcn/80500_en.html.

[12] Oksman, V., and J. Egan, (eds.), *Applications of ITU-T G.9960, ITU-T G.9961 Transceivers for Smart Grid Applications: Advanced Metering Infrastructure, Energy Management in the Home and Electric Vehicles,* Telecommunication Standardization Sector of International Telecommunication Union (ITU-T), 2010. Accessed March 24, 2016. https://www.itu.int/dms_pub/itu-t/opb/tut/T-TUT-HOME-2010-PDF-E.pdf.

[13] Latchman, H. A., et al., *Homeplug AV and IEEE 1901: A Handbook for PLC Designers and Users,* New York: Wiley-IEEE Press, 2013.

[14] HD-PLC Alliance, "Welcome to HD-PLC Alliance," 2015. Accessed March 24, 2016. http://www.hd-plc.org/modules/alliance/message.html.

[15] PRIME Alliance, "Alliance," 2013. Accessed March 24, 2016. http://www.prime-alliance.org/?page_id=2.

[16] Sendin, A., et al., "PRIME v1.4 Evolution: A Future Proof of Reality Beyond Metering," *Proc. 2014 IEEE International Conference on Smart Grid Communications,* Venice, Italy, No. 3-6, pp. 332–337.

[17] G3-PLC Alliance, "About Us." Accessed March 24, 2016. http://www.g3-plc.com/content/about-us.

[18] Private Mobile Radio, "PMR Technology," 2016. Accessed March 28, 2016. http://www.etsi.org/technologies-clusters/technologies/digital-mobile-radio/private-mobile-radio.

[19] ERC, *Methodology for the Assessment of PMR Systems in Terms of Spectrum Efficiency, Operation and Implementation (Report 52),* 1997. Accessed March 24, 2016. http://www.erodocdb.dk/docs/doc98/official/pdf/REP052.PDF.

[20] Dunlop, J., D. Girma, and J. Irvine, *Digital Mobile Communications and the TETRA System,* New York: Wiley, 1999.

[21] The IEEE 802.16 Working Group on Broadband Wireless Access Standards, "IEEE 802.16 Published Standards and Drafts," 2016. Accessed March 24, 2016. http://www.ieee802.org/16/published.html.

[22] ZigBee Alliance, "The Alliance," 2016. Accessed March 24, 2016. http://www.zigbee.org/zigbeealliance/.

[23] Wi-SUN, "The Alliance," 2016. Accessed March 28, 2016. https://www.wi-sun.org/

[24] Wi-Fi Alliance, "Who We Are," 2016. Accessed March 24, 2016. http://www.wi-fi.org/who-we-are.

[25] 3rd Generation Partnership Project (3GPP), "About 3GPP Home," 2016. Accessed March 27, 2016. http://www.3gpp.org/about-3gpp/about-3gpp.

[26] 3GPP2, "About 3GPP2." Accessed March 28, 2016. http://www.3gpp2.org/Public_html/Misc/AboutHome.cfm.

[27] CableLabs, "The Consortium," 2016. Accessed March 28, 2016. http://www.cablelabs.com/about-cablelabs/

[28]   Deployment & Operations Committee, "FTTH-Handbook - Edition 7," FTTH Council Europe, 2016. Accessed March 24, 2016. http://www.ftthcouncil.eu/documents/Publications/FTTH_Handbook_V7.pdf.

[29]   Broadband Forum, "The Forum," 2016. Accessed March 28, 2016. https://www.broadband-forum.org/

[30]   Fibre to the Home Council Global Alliance (FCGA), "Welcome to FCGA," 2015. Accessed March 24, 2016. http://www.ftthcouncil.info/index.html.

[31]   Commscope, "GPON – EPON Comparison," 2013. Accessed March 24, 2016. http://www.commscope.com/Docs/GPON_EPON_Comparison_WP-107286.pdf.

6

Telecommunication Architecture for the Smart Grid

The electric grid was conceived at a time where telecommunications had not yet been developed. As the challenge of the smart grid is the integration of information and communications technologies (ICTs) in general and the telecommunications in particular, this chapter presents alternatives to combine the different telecommunication technologies and networks into an architecture for the smart grid.

6.1 Network Architecture

6.1.1 Introduction

Power grids can be viewed as large infrastructures that have generation and transmission at their core. These are complex assets which require important, long-term investments and were not designed to become integral part of a telecommunication architecture or even to host telecommunication equipment [with the exception of the most modern primary substations (PSs) and some specific parts of the transmission system].

Beyond its "core" (generation and transmission), the grid is made of the "access" [i.e., medium voltage (MV) and low voltage (LV) distribution], which extends its reach up to every electric load in a given service area. Compared to generation and transmission, MV and LV assets require lower unitary investments and, again, were not thought to integrate telecommunication solutions.

One of the first decisions in the telecommunication architecture definition is a strategic one: a utility needs to understand how private telecommunication networks and public telecommunications service provider (TSP's) services are going to be combined. This decision will dramatically influence the architecture of the network such that, if utility-owned networks are to be favored, a more distributed architecture needs to be defined. For a smart grid mainly based on commercial TSP services, there will be generally fewer, more centralized interconnection points.

The knowledge of the grid is fundamental to leverage existing assets for telecommunication purposes, identifying electric grid premises, their location, and their nature, and deciding on the role they could play in the smart grid network. For example, if secondary substations (SSs) are the sites where service devices [data concentrators (DCs), intelligent electronic devices (IEDs), and so forth] are to be installed, and PSs are the natural locations where traffic from several SSs could be aggregated, both elements play a role in the telecommunication network and should be considered part of it.

Independent from the strategy (e.g., private or public networks combination) and the technical options, Internet Protocol (IP) will be the network layer protocol to be used. All industries and markets currently offer a full range of devices and components that leverage the ubiquity of IP to offer lower-cost hardware and software.

6.1.2 Architecture Blocks

Telecommunication networks can be segmented and classified according to the environment or extension area of the network. Acronyms like WAN, FAN, NAN, MAN, HAN, WLAN, LAN, PAN, GAN, CAN, and SAN are common nowadays. No agreed definition exists for many of them, and a simpler approach will be proposed for the smart grid.

The function of the network architecture is to take the traffic from the sources in the service points (substations, meters, and so forth) to the service delivery points where the application servers will be located in a scalable, manageable, and efficient way. The following blocks are proposed for the smart grid architecture:

- Core network: It is the central block of the architecture, the part of the network where all elements can be interconnected and their traffic carried to destination. The core network consists of the most important telecommunication nodes as any failure will have major effects on the whole network; in this regard, it is similar to the high voltage (HV) grid. Meshed connectivity topologies, high capacity and reliability, maximum performance, and scalability are usually associated to the core network. Thus, gigabit-capacity transport technologies (e.g., wavelength division multiplexing (WDM), synchronous digital hierachy (SDH), synchronous optical network (SONET) are used to interconnect relevant network aggregation points, capable of routing high-volume traffic. Depending on the total network dimension, the core network can have from a few to hundreds of routing elements.

- Access network: It is the part of the network used to reach distribution grid elements (i.e., substations). Most of the elements in this access network are within the MV grid. The access network will generally include different technologies. As final service locations will be spread all over the territory, different technologies will be optimal (technically and economically) for each specific case. Public access networks [2G/3G/4G, xDSL, high fiber coaxial (HFC), and so forth] or different private network solutions [private radio, broadband power line (BPL), optical fiber] will connect MV elements to the core network. SSs are one of the most relevant and numerous elements to be connected through this access network.

- Local network: It comprises the telecommunication network between the SSs and the electricity meters, approximately corresponding to the LV grid segment. This LV segment (and accordingly the local network) has more or less relevance in the different world regions and in the different parts of each grid (in the United States it might be negligible in areas where LV customers are connected one-to-one to the MV/LV transformers; however, in Europe it is an important part of the grid, although in some areas the number of customers per transformer could be low). Local networks will not be further discussed in this chapter since their main components are smart meters, which usually do not need specific network planning or design; the reason is with the technologies used (generally of a mesh and self-configurable nature) and with the scale of the area (small) where they are deployed before they connect to the access network. Chapters 7 and 8 will provide further details on this topic. However, local networks will become increasingly important once MV grids are fully controlled by smart grids, as LV grids (where applicable) will then be the next frontier in the grid and will progressively require similar operational procedures to those of MV grids. If and when that happens, network planning will be needed for local networks as well to optimally integrate new LV control devices into the smart grid.

Transport and packet switching/routing technologies are present in each of the architecture blocks above. Transport offers the capability to structure the transport of data mainly where long distances are involved; packet switching/routing is needed to aggregate and drive the traffic to its destination. Figure 6.1 shows an example of how the architecture is structured. The core network

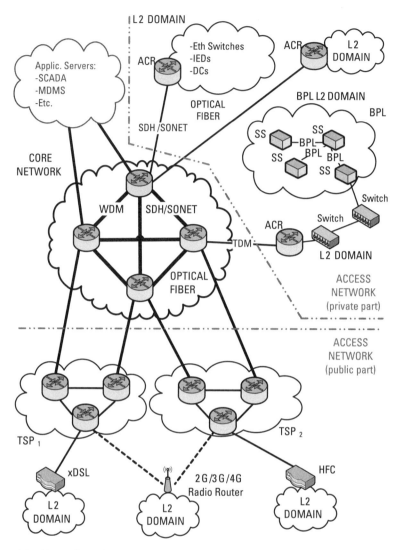

Figure 6.1 Network topology.

uses different high-capacity transport technologies to connect core routers. Core routers are installed at service centers, PSs, and even relevant SSs. The access network aggregates different access clusters (groups of elements connected together in an area of the access network) through access connection routers (ACRs) onto the core routers. Some of these clusters will aggregate services, with switches, at an SS, which, in turn, will connect via, for example, BPL to other substations such that one of them will have a transport connection (be it private or public) to the core network. Some other clusters will use direct public network connectivity (e.g., xDSL, HFC, 2G/3G/4G radio).

Local networks are basically groups of smart meters connected with power line communication (PLC) or radio (be it private or public).

6.1.3 IP Network

IP network planning and design are instrumental across both the core and access networks, and probably in the future within the local network. IP design affects overall network performance and scalability, and it must guarantee that network operation is manageable. A flexible IP plan that can adapt to future needs in the evolutionary scenario of smart grids is fundamental.

A key aspect of IP network planning is IP address allocation. IP segmentation needs to be adapted to the existing and future utility needs while ensuring an efficient use of the limited IP addressing space. Smart grid services usually require private IP addresses as, even when public commercial services are used, it is not common to associate public IP addresses to any utility service device or terminal. IPv4 private ranges are normally enough to cover smart grid deployments if local networks are not considered part of the IP planning. IPv6 is an alternative option if the addressing space needs to be enlarged, but the level of adoption of IPv6 does not make IPv6 a common deployment practice for smart grid networks today.

Depending on the network block, and segment, IP packet routing (L3) or Ethernet frame switching (L2) can be used. A combination of L2 and L3 domains must be defined to design the most efficient and manageable data network. While L2 network setup and configuration is simple and cost-effective, L3 improves troubleshooting operations, scalability, and redundancy implementation:

- Layer 3 centralized configuration: Routers with IP packet routing capabilities aggregate traffic from distant locations connected at L2 among them (Ethernet switching). For the sake of simplicity, we will refer to this configuration as L2.
- Layer 3 distributed configuration: Routers are placed close to the endpoints (hosts). All the premises connected in the network edge should have a router inside. For the sake of simplicity, we will refer to this configuration as L3.

The combination of these configurations depends on the particularities of each deployment. The first step to determine where to configure L2 and L3 domains is the selection of the sites where the telecommunication solution will be considered an extension of a L2 domain. Every substation can be seen as a single L2 domain, but depending on the access technology (e.g., BPL, carrier

Ethernet) the L2 segment can be extended to additional substations. However, even in the same grid, different substations may have different requirements and hierarchical relevance, and this justifies having different telecommunication solutions for them. Thus, there will be substations designed with dedicated routers (L3) and other substations will be deployed with switches (L2) connecting to other substations. This L2 concentration, extended through several substations, will be connected to an ACR (L3) at the point of aggregation to the core network. This structure must be supported by the IP addressing plan, as the subnetworks to be created (with subnet mask) must be large enough to host the elements in the different concentrations.

L2 domains organization is another aspect to be considered. In Ethernet L2 domains, every device can become part of the network by itself and discover other nodes using the ARP (as in RFC 826). This makes it easy to aggregate new elements to the domain, and thanks to the use of virtual local area networks (VLANs), as in IEEE 802.1Q, a logical segmentation within a local area network (LAN), the network can be organized in different logical areas needed for each service. IP planning is obviously affected by this. Normally, each service requires only an IP subnet for each LAN/VLAN; however, sometimes it may be helpful to use two separate IP subnets in the same LAN/VLANs for the same service if there is a need to have a primary and a secondary IP address for path protection purposes.

Once the services coexisting in the L2 aggregation are defined, the list of LANs managed by the same router should be defined. This router acts as a default gateway [1] for these services, and will route the traffic to and from application servers. A minimum of one LAN per service is defined; an extra number of LANs should be reserved to guarantee the connection of new services for the evolution of the smart grid. The size of each LAN needs to be calculated to define the subnet masks; the number of substations and devices sharing the same LAN should be considered.

The scalability of the L2/L3 architecture needs to consider the total number of substations to be connected. If an L2 aggregation configuration is going to be used to group different services from several locations, the percentage of the substations to be connected directly with L3 solutions and the percentage that is aggregated through L2 connectivity needs to be defined. In order to efficiently handle these large interconnected networks, the concept of route summarization, that is, aggregation or supernetting (RFC 4632), must be applied, to divide a large network into smaller ones that are assigned to the final locations that are managed by the same router. With this strategy, when the router is taking decisions about the next hop in the route, the entry in the routing table is aggregated into a larger network that is the combination of the smaller subnetworks; routing tables get simplified and network resources optimized. This is one of the key goals of design and planning for network scalability;

subnetting/supernetting strategies according to service delivery and geographic criteria ease traffic management in distributed architectures.

The core network needs its own L3 IP definition. Depending on the chosen technology, the definition of IP addresses for the network core connectivity can be different. In the case of private networks, this can be solved with the definition of the IP subnetworks to route the packets in the core routers. However, with public service solutions, an IP addressing that is compatible with the TSP IP network is needed; tunneling strategies can be used [e.g., IPsec as in RFC 4301, Generic Routing Encapsulation (GRE) as in RFC 1701 and 1702)].

6.1.4 Isolation of Data Flows

Visibility of traffics in the different parts of a network and its services needs to be controlled in a multiservice telecommunication network. Data flow separation is important in a context where a single telecommunication network is shared by independent users and applications.

Data flows are kept separated with L2 and L3 policies. These policies must be flexible enough to allow modifications as the smart grid evolves. For example, networks need to be flexible to dynamically activate or deactivate the visibility of different substations within the same service, without the traffic being routed up to any central application server; this could be useful to allow the transition from centralized applications (e.g., remote control from a central application), to a distribution automation (DA) in which substations need to interact through the telecommunication network directly.

QoS is also a key concept in the provision of critical services through shared network resources. Once the data flows are differentiated, traffic prioritization can be applied to achieve the needed performance (throughput, latency, and so forth). There are several techniques at different layers to identify traffic which has to be prioritized and different policies to be applied when traffic priority is identified based on L2 or L3 properties. For example, the type of service (TOS) field of IP datagrams can be used to assign different priorities depending on its value (further details about QoS techniques, implementation, and evolution are in [2]).

Depending on the particularities of each utility (internal responsibilities assignment, type of smart grid services used, telecommunication network and grid operation, and so forth), the policies to apply for each traffic flow will be different; the delivery points of each service may also be different. Generally, MV automation will be managed by a different application than smart metering; thus, service delivery points will be different. Substation surveillance services may be delivered to specific control centers, which can actually be distributed in several geographical locations. Traffic from a PS will be more critical than traffic from a SS and, with regard to smart metering traffic, connect and disconnect commands will be more critical than an instantaneous meter reading. Traffic categorization can be as complex as the telecommunication network may support, but needs to be just as simple as required to have a network that is easy to operate and maintain.

Policies used for traffic prioritization must be coherent in every block of the network architecture for each particular traffic flow from the access network, through the core network, to the delivery of traffic at the service delivery points where the central application servers are located. Prioritization of data flows needs to be kept end-to-end and is often a challenge for certain TSP services (e.g., radio access).

6.2   Access Network

The access network is made of telecommunication elements connecting SSs (and other MV elements) among them and into the core network. Data traffic circulates between end devices and central application servers. The access network must support traffic aggregation while guaranteeing end-to-end performance with a coherent design across the entire network (including the core network).

In order to establish the telecommunication needs for this access network, it is necessary to identify both the services to be aggregated at each point of the network and the individual requirements for each. Because not all services have the same priority, the network needs to support QoS for different traffic flows as part of the design.

These inputs, together with the economic considerations and the strategic vision of the company, have an impact on the technology decisions to be taken. The advantages and disadvantages of private networks and public services options will have to be evaluated and an adequate combination of both selected.

Within the private networks domain, there are point-to-point technologies such as optical fiber or radio links (with different alternatives, depending on the performance needs) that will be useful for transport purposes; technologies oriented to point-to-multipoint scenarios such as private mobile radio (PMR) data networks and digital radio systems (WiMAX, Wi-Fi, and so forth); BPL in point-to-point or point-to-multipoint configurations will efficiently connect secondary substations over MV power lines.

Among public services provided by TSPs, the very generalized availability of broadband connectivity makes it possible to use wireline xDSL, HFC and FTTx services, and wireless services such as second generation (2G), third generation (3G), and fourth generation (4G) at selected SSs.

6.2.1 Telecommunication Service Requirements

Applications and underlying telecommunications are interdependent. It is not reasonable to design an application under the assumption that the telecommunication network has always the necessary capacity, and similarly, it is not realistic to think that applications will remain constrained by the limitations of given telecommunication networks. Thus, both applications and telecommunications need to be designed considering one another. In particular, applications and their systems need to take into account the possible limitations of the telecommunication technologies that could exist (throughput, latency, jitter, and so forth). For the same reason, if an application forces a mandatory telecommunication requirement, the telecommunication network or TSP service needs to support it, if it is technically and economically feasible.

The new applications for smart grids (see Chapter 4) impose more requirements for telecommunication networks and services than traditional applications. These requirements are stricter both in terms of the services themselves (e.g., better throughput, latency, jitter) and also in terms of larger service areas, covering HV, MV, and even LV assets. Traditional telecommunication technologies need to adapt to these new requirements. Secure, managed, two-way digital telecommunication services are to be provided in a context of broadband traffic and automatic redundancy.

Table 6.1 summarizes some commonly accepted requirements for the new smart grid services in the access domain (distribution grid). Some utilities may use more stringent requirements in the services or subservices within the main service.

There are other important associated network traffic-related aspects not covered in Table 6.1:

Table 6.1
New Smart Grid Services Requirements in the Access Network

Service	Bandwidth (kbps)	Latency	Availability (%)	Security	Power Supply Backup
Advanced metering infrastructure (AMI)	10–100	2–15 sec	99–99.99	High	Not necessary
Distribution automation (DA)	9.6–100	100 ms–2 sec	99–99.999	High	24–72 hours
Demand response (DR)	14–100	500 ms–several minutes	99–99.99	High	Not necessary
Distributed generation (DG)	9.6–56	20 ms–15 sec	99–99.99	High	1 hour
Electric vehicle (EV)	9.6–56	2 sec–5 min	99–99.99	Relatively High	Not necessary

Source: [3, 4].

- Use of the services:
  - Electronic devices (DCs, IEDs, etc.) usually provide a command line interface (CLI) and/or a graphical user interface (GUI) for configuration and operation. A CLI usually demands little bandwidth, but a GUI often relies on an internal Web server that is accessible through http protocols, and depending on its design sometimes excessive bandwidth is demanded from the network connection. The traffic pattern for GUI access (which is manual, on-demand, and sometimes done in parallel over several devices) and the regular, automated access from central systems are drastically different.
  - Firmware upgrade operation is very common in microprocessor-based electronic devices. This type of management traffic can be heavy, depending on the size of the firmware image and the protocol used in the process. The simultaneous download on hundreds or thousands of devices can collapse the network or the interconnection with third-party systems, if not properly scheduled or dimensioned.
- Network overhead:
  - Traffic and protocol overhead. Telecommunication protocols include certain additional information that is added to the service payload (e.g., information on packet origin and destination, priority, classification, security); moreover, not only the overhead exists in terms of additional data, but also as an increased number of transactions (e.g., registration and authentication procedures, acknowledging). This information is not associated to the service itself but to the network, and it means that the traffic in the network is heavier than just the service payload. Calculations made with short length payload packets (e.g., DA traffic) show that the overhead may be larger than the payload itself.
  - Management overhead. Any telecommunication network needs to be managed, so this could mean additional management traffic in the network. Traditional transport protocols natively include management overhead (e.g., SDH/SONET) to be handled by network management systems (NMSs); however, the end-to-end connectivity in TCP/IP networks may need and extra amount of traffic, additional to useful traffic [e.g., Internet Control Message Protocol (ICMP) polling]. Furthermore, it has to be considered that not only the network has to be managed, but also the service elements (e.g., DCs, IEDs).

It is relevant to note that out of the different telecommunication service requirements there is usually one which will be the most constraining one. This constraining telecommunication service requirement needs to be considered the initial driver of the network design. This requirement will limit the applicability of certain technologies, and will make the rest of the requirements not so immediately relevant for the design.

Service requirements establish a theoretical starting point, but they cannot often be satisfied in all circumstances. Ranges with minimum and maximum acceptable thresholds are usually defined for the most relevant parameters, so that state-of-the-art networks and services can be used. Performance compromises should be agreed, and applications have to accept some operational limitations. Examples of these constraints are:

- Simultaneous use of a dedicated link by different applications. Different applications may need to be coordinated in the remote access to a single end-point, to avoid saturation of a low-speed telecommunication link that may not be able to prioritize data flows.

- Simultaneous use of a network link by the traffic to different end-points. Network links dimensioned to the normal conditions highest usage (peak) may not be ready to absorb multiple parallel unexpected data streams or the collapse of other network links if high-capacity, demanding applications over a multiplicity of end-points would be required to work simultaneously.

6.2.2 Design Alternatives

Table 6.2 shows a brief summary of some technologies commonly used for smart grid in the access network. Some simplifications have been made (e.g., some technologies can be supported over different media and the most common or representative case has been selected). The advantages and disadvantages are focused on the network in the private infrastructure cases, and on the services in the case of public commercial networks.

Table 6.2 options are not easily comparable. Private network options imply higher investment costs that need to be compared with the recurring costs of the commercial services alternatives. However, commercial services often present performance and QoS limitations with respect to private network options. Wireless technologies are often easier to deploy than wireline alternatives. However, where the infrastructure already exists (e.g., xDSL, HFC, power line, optical fiber), wireline approaches become an attractive solution in terms of time to deploy, performance, and reliability. However, access to spectrum to deploy private wireless solutions may be in practice impossible, so the use of

Table 6.2
Telecommunication Technologies for the Access Network

Technology	Ownership Model	Physical Medium	Advantages	Disadvantages
Point-to-point TDM/Ethernet over TDM	Private/commercial	Various	Apparent simplicity, network security	Not scalable for network deployments, TDM is in obsolescence process
Frame relay	Private/commercial	Various	Adapted to TDM services	High cost, not scalable in terms of throughput, FR is in obsolescence process
Carrier Ethernet	Private/commercial	Optical fiber	Native Ethernet, native VLAN segmentation, data rate scalability, cost of interfaces	Not available in a generalized way
IP/MPLS	Private/commercial	Optical fiber	L2 and L3 services, high security, high performance	The utility end-points must be connected to the MPLS core usually over a dedicated point-to-point connection
MPLS-TP	Private	Optical fiber	Reduced set of MPLS, simplified operations.	Transport oriented
BPL	Private	Electric grid	No extra wires needed, reduced cost compared to alternative private options	Coupling to power lines needed, not applicable to the 100% of the power lines (distances and cable status)
Cable services (HFC)	Commercial	Optical fiber & Coaxial cables	High penetration in dense urban areas, high performance	Not always available close to every SS, terminal equipment not commonly ready for industrial environments
Digital subscriber line (xDSL)	Private/commercial	Twisted pairs	High penetration in dense urban areas, high performance, copper pairs are available in substations in some countries	Not always available close to every SS, terminal equipment not commonly ready for industrial and similar environments
FTTx	Private/commercial	Optical fiber	High capacity	High cost, not available in a generalized way
2G (GPRS/EDGE)	Commercial	Wireless	High penetration, easy service access	Low performance, recurring cost usually based on traffic, no traffic guarantee possible, no network access guarantee possible, coverage difficulties in rural and underground premises

Table 6.2 (continued)

Technology	Ownership Model	Physical Medium	Advantages	Disadvantages
3G (CDMA 2000/WCDMA/HSPA Rel. 5/6/7)	Commercial	Wireless	High/medium penetration, easy service access	Network designed for smart phone-like traffic, recurring cost usually based on traffic, no traffic guarantee possible, no network access guarantee possible, coverage difficulties in rural and underground premises, coverage may be affected by concurrent users in the area
TSP 4G LTE	Commercial	Wireless	Low penetration, high performance, IP and QoS natively available	Network designed for smart phone-like traffic, recurring cost usually based on traffic, coverage difficulties in underground premises
Utility (private) 4G LTE	Private	Wireless	High performance, IP and QoS natively available	Spectrum access (availability and costs), high cost of reaching close to 100% coverage
Satellite telecommunications	Commercial	Wireless	Available, typically independent from terrestrial networks	High cost, high environmental and mechanical constraints (big antennas), propagation delays, coverage difficulties in underground and urban areas
RF mesh over unlicensed frequency bands	Private	Wireless	Low device costs, extensible coverage with repeaters	High number of repeaters (and their infrastructure) needed, low throughput, high latencies, no QoS, subject to interference
WiMAX (licensed or unlicensed frequency bands)	Private/commercial	Wireless	Pure end-to-end IP connectivity, short- and long-range access, symmetrical bandwidth	Spectrum access (availability and costs), line of sight (LOS) usually needed
Wi-Fi (unlicensed frequency bands)	Private	Wireless	Low cost of interfaces	Local coverage, limited accessible spectrum, subject to interference

commercial services or the usage of unlicensed spectrum, which may suffer from interference, become the alternative.

6.2.2.1 Wireline Versus Wireless

One of the natural benefits of wireless (radio) technologies is mobility; in order to provide mobile telecommunication services a coverage area needs to be generated. Thus, even if fixed premises associated to the smart grid will not take advantage of the mobility itself, they can benefit from the preexisting coverage areas.

The installation and commissioning of wireless solutions are more flexible than their wireline counterparts. On the contrary, wireless systems depend on environmental aspects, which, if anticipated at planning and design phases, can only be statistically forecast. Multipath, fading, and moving obstacles cannot be avoided so unexpected conditions may arise and will manifest themselves as unavailability and performance limitations.

For end-points to be part of the coverage area, antennas are needed. Antennas and, in general, the radiating system (consisting of duplexers, cables, waveguides, in the case of microwave links, and auxiliary elements) need to be carefully considered:

- Antennas need to be installed in the best possible location with the highest permanently received signal (difficult in urban scenarios where the signal is affected by multipath and transmission from different radio stations).

- Antennas are best placed outdoors. However, both the environmental and social impact of these elements make the installation and surrounding imitation process challenging.

- Antenna selection and mounting conditions have to respect the available space (antenna size is fully dependent on the frequency; ground plane, if needed, cannot be neglected).

A major concern with wireless networks is how the radio access is defined and implemented. In most common radio systems, the access segment cannot guarantee the access of some of the users of the system in detriment of others. Even if at some of the parts of the system QoS can be managed, the access to the system cannot offer an end-to-end continuation of the QoS.

The flexibility of wireless systems is their main advantage. However, the higher predictability of wireline technologies recommends their use where possible for the access network. The mix of wireline and wireless networks will be influenced by the factors already mentioned, plus others such as the leveraging of already existing networks, the size and geographical extension to be covered, the distribution of assets in urban, suburban, and rural areas, their disposition type (above or underground), and the available TSP services.

6.2.2.2 Narrowband Versus Broadband

Absent any other consideration, broadband telecommunication technologies should be the preferred option for the smart grid access network, as smart grid represents an evolving concept open to new services and requirements. Thus, in general, the broadest the band reaches the end-points, apparently, the better.

However, narrowband technologies cannot be dismissed for smart grids because many of their associated services, individually taken, could be provided with narrowband telecommunication networks. For example, MV reclosers in automation contexts could just be served with a narrowband (private or public) radio-based technology; this could be the case at rural and dispersedly populated areas with small SSs (without any short-term need for advanced services). The common drawback of narrowband technologies is the usual (with few exceptions) absence of real-time management capabilities (due to their narrowband nature and/or public service origin); this is why they cannot be generally recommended for a broad base of critical smart grid services.

The discussion of narrowband versus broadband should be replaced by that of single-service versus multiservice networks, where a narrowband solution is very limited in the latter case. Multiservice networks are recommended due to their cost savings in terms of installation, operation, and maintenance. If single-service networks are to be used, they need to be integrated within the same operational framework as the rest of utility access network components.

6.2.2.3 Private Networks Versus Public Network Services

Chapter 1 already introduced the comparison between private networks and public services provided by TSPs for smart grids. Historically, many utilities have been using private telecommunication networks to support core businesses when considered critical services.

TSPs deploy networks to run a commercial business targeting a certain customer range. TSP networks are designed and developed for that customer base and intended geographical area, with the QoS needed for the expected traffic. If the TSP is successful, it will get the expected customer base and the network will perform correctly. If the TSP is too successful, it may happen that the traffic demand by its customers is such that the TSP network underperforms. In a competitive market, customers may be unsatisfied while the network grows to offer all of them a good QoS; this translates into customers moving among different TSPs while the network consolidates. This legitimate business situation may result in adverse scenarios for utilities with critical service demands if TSPs do not provide mechanisms to adapt existing networks to smart grid utility requirements.

In terms of traffic pattern adaptation, most existing public mobile networks assume Internet-like traffic with downstream being prevalent; this traffic is being typically demanded by Web browsers that behave with a certain pattern that is used by the network to assign its resources. Another typical situation is the oversubscription of networks, where statistically TSPs consider they can support additional services because not all users will be constantly demanding the contracted traffic. Traffic patterns that do not comply with those behavioral standards may go underserved. As a consequence, public networks do not necessarily fit well with professional service delivery and mission-critical applications, particularly in wireless access networks.

The challenges associated with public network services also include the possible changes implemented on those networks. Any network needs to evolve to adapt to technology changes, obsolescence of components and license conditions such as frequency plan modifications. These changes are out of the control of the customer, and utilities realize that, if the services are provided through public networks, network changes may affect them. Utility smart grid business plans need to anticipate these contingencies.

In any case, public networks that are already deployed over vast geographic areas represent a huge advantage for the utility and allow balanced approaches in terms of private network investment and use of public services.

To secure private network investments, utilities need to identify the best network technologies for their specific case, considering their life cycle, external dependencies, and existing utility infrastructure. Life cycle is important because investments associated with private network deployments need long amortization periods that have to take obsolescence into account. External dependencies are critical, for example, a consistent number of alternative vendors need to exist in an ecosystem that guarantees the evolution of the technology; more dependencies are created around frequency licenses that are assigned for a certain period of time. Existing utility infrastructure is fundamental not only to enable the business plan, but to facilitate network development as well.

6.2.2.4   Transport and Packet-Switching Networks

Telecommunication technologies classify themselves around some broad categories shown in Chapters 2 and 5. Based on that, it may be suggested that transport and packet networks are antagonistic technologies; however, they happen to be complementary technologies with a progressively blurring border.

Many utilities' networks started to be deployed at a time when telecommunications were just a matter of carrying flows between two points using pure transport. When packet networks started to be broadly used, if any service (including IP) on a remote location had to be connected with the central application servers, the transport network offered a service access point, transparent to L2 or L3 protocols, to carry the data to the delivery point. This basic architecture was a consequence of legacy serial protocols and forced the applications to manage anything above L2.

This kind of strategy is not scalable. It naturally evolves to include L2 services in the telecommunications network, optimizing its use. In this scenario, the final services may find a switch at remote sites, enabling the aggregation of services through the use of L2 protocols to connect with the central systems, where routers and servers are located. This approach is more scalable, but it may generate difficulties when new services and large number of devices are present in the network, due to the broadcast nature of L2 protocols. The need for domain segmentation arises due to the need to implement a better traffic control and bandwidth management. These features are better managed using L3 networks closer to the end-point.

Each utility, depending on the size of its grid, its services' requirements and its stage in the evolution towards smart grids, may use transport and packet switching in one way or the other. However, the evolution from simple point-to-point transport, L2 switching, and L3 routing are concepts to be managed. The use of IP, as already defined, implies that the access network nodes must have IP traffic aggregation capabilities. Basic transport would be an option in very specific and limited situations. Switching is needed whenever IP traffic needs to be aggregated and is a simple way to make traffic progress; however, scalability and troubleshooting limitations make it applicable just to some segments of the network. Routing is the scalable option, but careful IP design is a must in a network where the services are in constant evolution. The final scenario is clearly one where each SS has transport capabilities to take the traffic from a switch or a router to the rest of the access network.

The best way to combine L2 and L3 services (i.e., SS networking based on switches versus routers) has no immediate answer. Historically, external factors have forced L3 devices provisioning at the end-points, as a result of the TSP's available connectivity. TSP services based on radio networks are not available at L2, and only recently L2 alternatives have been offered as services over certain wireline networks (e.g., optical fiber).

On the contrary, if the SS is connected through a private network, the extension of L2 domains between SSs or even PSs is a good solution. The traffic can be aggregated at L2, and L3 devices can be located at aggregation points closer to the core network. The connectivity of these L3 devices can be made with pure transport technologies. The use of switches (L2) provides a simpler (in terms of commissioning and operation) and more cost-effective solution than the use of routers (L3), in which the aggregation capacity of Ethernet ports is lower.

Figure 6.2 shows the combination of the three approaches. In the example, one SS aggregates services in a switch and the traffic is progressed through a BPL link acting as a pure bridge. At another SS, there is an optical switch that

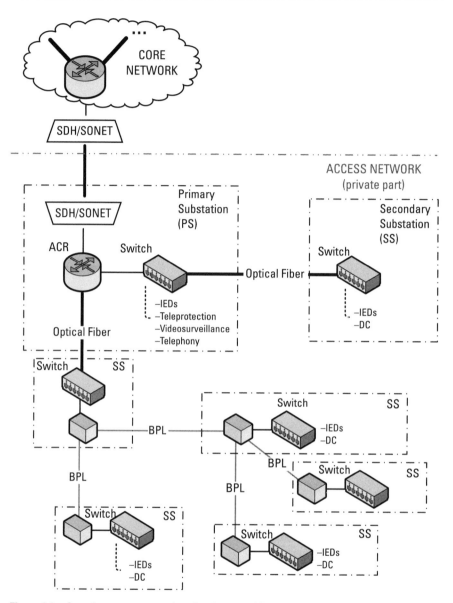

Figure 6.2   Sample access network technology combination.

transports the L2 domain to the next hop placed in a PS. In the PS there is a router interconnecting services from different SSs, together with its own services. This ACR has an interface connected through a SDH/SONET transport device to a core router located in the core network.

6.2.2.5 Aggregation of Services in Electrical Grid Sites

Typical network traffic aggregation points (Figure 6.3) are:

- Secondary substations: Specific SSs can aggregate traffic from other SSs (e.g., connected through optical fiber or BPL) additional to their own traffic. The expected size (number of SSs and services) of this aggregation is limited and it may be implemented at L2 (i.e., based on Ethernet switches). In this scenario it is important to consider which the rules for visibility between devices and services will be. A possible approach is

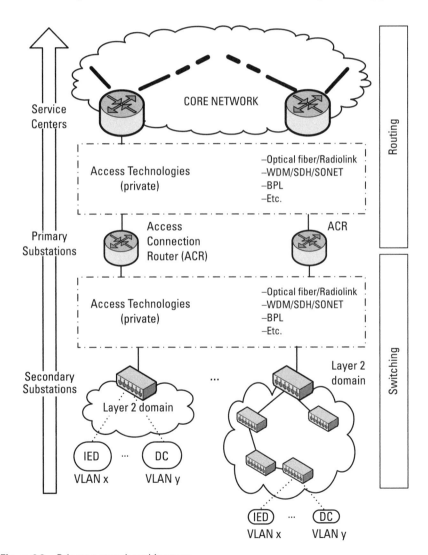

Figure 6.3 Private network architecture.

to have a dedicated VLAN for each service at every SS L2 aggregation. However, this implies a specific configuration for each L2 device in the network, and accordingly a daunting configuration workload in the case of massive deployments with thousands of devices. An interesting alternative is the reuse of the same VLAN on each SS to identify a service, acknowledging that it will not be easy to limit visibility among devices which share the same VLAN (it may be solved using private VLANs, as proposed by RFC 5517).

- Primary substations: Traffic from every SS connected to a PS is aggregated onto it. An ACR is generally installed to group all the traffic and send it to the core network. Typically, all the LANs managed by such a router will be announced by a routing protocol.

- Service centers: The aggregation of traffic from several PSs in an area can be also performed in nonelectrical premises that can be used as core network router sites.

6.3   Core Network

The core network consists of the group of core routing elements and the transport to connect them when distances require it. These core routers aggregate and route all the traffic of the network with the specific policies required for each data flow. These core routers can be implemented as physical or virtual routers (e.g., if provided through an MPLS platform).

The objective of the core network is to interconnect all remote end-points and with the application servers. The internal core network architecture must allow traffic disaggregation from the different services in order to separately manage each data flow. This management will depend on the type of traffic (e.g., depending on the application protocol), on the type of service (e.g., DA, AMI, video surveillance), and geographical region.

MPLS is the technology that supports packet switching functions of the core network in an optimal and flexible way. MPLS provides a packet traffic forwarding platform combining the advantages of L2 (speed and simplicity) and L3 (routing control) technologies (see Chapter 5). MPLS allows traffic engineering, isolation of data flows, fast rerouting, and mechanisms for network security, granting high availability and appropriate performance. TSPs all over the world have adopted MPLS as the technology for their core networks, and many utilities too as they develop private core networks for smart grid services. The advantages of MPLS services for the smart grid telecommunication network can be summarized as follows:

- Configuration of multiple simultaneous virtual private networks (VPNs; it is a private data network within a common telecommunication infrastructure) at L2 or L3:
  - Virtual pseudo-wire service (VPWS), as in RFC 4447, known also as VLL service. A pseudo-wire is point-to-point connection between two end-points.
  - Virtual private LAN service (VPLS), as in RFC 4762. A VPLS is a L2 VPN service used to provide multipoint-to-multipoint Ethernet connectivity.
  - Virtual private routing network (VPRN) service, known as MPLS VPN based on RFC 4364. It provides L3 connectivity among end-points.
- Sharing of network infrastructure, maintaining logical separation and QoS.
- Efficient traffic management due to multiple data streams sharing a common transmission path.
- Rapid protection against failures, at the network link or node level.
- Highly efficient traffic switching and routing, based on simple MPLS labels.
- Independency between data and control planes.

6.3.1 Telecommunication Service Requirements

The core network needs to support the requirements of all applications (see Table 6.1) deployed at remote sites. Thus, it has to be engineered to match the design criteria of the access network. In essence, the core network needs to integrate the access network as an extension.

All service requirements should be considered during the design stage. The most obvious requirements or the most demanding applications tend to focus the attention and make other relevant requirements fall behind. Thus, a network could become functional and not be compliant with some of the aspects that are required. Some typical aspects that are often neglected are physical redundancies of network nodes and links, insufficient power supply backup, cybersecurity, and scalability.

A special mention is needed to network and service operation and management. Network performance must be guaranteed, and this can only be accomplished if an appropriate NMS exists to operate the network. Network operation includes the commissioning of the services, but also the network design assurance. NMSs must allow proper problem detection and troubleshooting, and must support network performance monitoring and trend analysis.

6.3.2 Design Alternatives

Network design is an iterative process, as not all data is available since the beginning. This is more obvious in the context of new networks with moving requirements, as it is the case of smart grids. Moreover, if the telecommunication network is built alongside with the development of the new services and applications, the hypotheses that have been fixed at this design phase might need to be revisited as the applications become operational. Typical assumptions that may make a design evolve are the unexpected traffic patterns for services and new use cases for certain applications (e.g., intensive monitoring needed for IEDs; firmware upgrades that need to be performed more frequently or over a larger number of devices than expected).

There are some mandatory inputs that are fundamental to make the right design choices:

- Performance, reliability, resilience, scalability, and security requirements of the different services.
- Specific particular requirements for selected services [e.g., Mean Opinion Score (MOS) in Voice over Internet Protocol (VoIP)].
- Expected traffic at every end-point of the network.
- Limitations of the existing private telecommunication network of the utility.
- Limitations of the available commercial telecommunication services in the utility area.
- Dependencies with network proprietary solutions. The use of standard protocols that assure interoperability among vendors is mandatory for each network solution.
- Operation and maintenance easiness, by means of simple configurations and management tools.
- Budget allowance.

The traffic from the end-points through the access network is aggregated in traffic concentration nodes (ACRs). ACRs connect to the core network nodes, located in the places where they can better perform their function, considering:

- Connectivity with ACRs.

- Internode telecommunication link (transport) and routing capacities of the telecommunication devices (within the architecture and the protocols used).

- Reliability aspects, with power supply, backup and diverse routes between each element and the network (at least two different destination nodes).

- Physical and cyber security aspects (accesses, sensors, firewalls, IDS systems, and so forth).

After the physical aspects have been drafted, the logical design must follow. The logical design needs to take into account the physical relationship among the nodes, and it should include:

- The logical connectivity among nodes;
- The networking protocols;
- The routing details;
- How redundancy is implemented with the selected protocols;
- QoS policies;
- Security policies.

Design processes cannot be executed without the help of network design tools. Network design tools help to qualify different design alternatives (scenarios) [5]. The tools usually have libraries of devices, protocols and link models, with adjustable attributes (traffic load, device throughput, and so forth), to simulate performance while showing traffic flows and highlighting bottlenecks. Network congestion, network link and node loads, response times, and failure and error conditions are included as study objects. However, real proofs of concept, laboratory tests, and small scale trials are needed to verify final designs.

In the occasions in which a pre-smart grid telecommunication network exists, a few practical ideas can be given to evaluate if the existing core network must be replaced:

- Determine pre-smart grid baseline traffic, and growth trends.

- Incorporate the new design requirements associated to the smart grid deployment, to create a long-term traffic evolution forecast.

- Optimize the existing core network design trying to leverage existing spare capacity from the existing network.

- Evaluate next-generation technologies alternatives to obtain a comparable network to the legacy core.

- Compare the pre-smart grid-evolved network to the next-generation green-field alternative, in order to understand which one is the most technically and cost efficient option.

6.3.2.1   Integration of Public Network Services in the Core Network

Private networks are often not enough to provide all services in all the service territory. When private network assets are not available or are not sufficient, TSPs become an alternative, for example, by means of leased lines. The services that TSPs can provide are generally well fit for the smart grid requirements in the core network. In fact, service-level agreements (SLAs) are clear and controlled rigorously. Table 6.3 shows a subset of typical TSP's services for core networks.

6.3.2.2   Delivery of Public Network Services in the Core Network

Public networks access traffic must be also aggregated into the core network. Traffic coming from wireless or wireline networks (see Figure 6.4) is usually delivered at L3 at interconnection locations agreed with the utility. The integration of these traffic flows in the architecture is to be agnostic from the kind of

Table 6.3

TSP's Services for the Core Network

TSP Service	OSI Layer	Description
Leased TDM Lines	L1	The links are dedicated for the traffic of the utility. Bandwidth and latency are predictable. L2 and L3 operation are responsibility of the utility.
Frame Relay	L2	Allows oversubscription of data services (circuit oriented), controlled by CIR and EIR. Cheaper than leased lines, but with risk of bandwidth oversubscription. The utility has to configure IP over FR and provide L3 services (routing capabilities). TSPs are phasing out this service, being usually replaced by MPLS services.
Carrier Ethernet Service	L2	Dedicated point-to-point or point-to-multipoint Ethernet connections between utility premises. It extends the Ethernet paradigm through the WAN. The utility has to configure IP over Ethernet and provide L3 services (routing capabilities). Not always available.
MPLS VPLS Service	L2	The TSP offers a geographically distributed Ethernet broadcast domain. All nodes connect to the VPLS cloud over Ethernet, optical fiber, TDM lines, and so forth. The utility has to configure IP over Ethernet for the WAN and provide L3 services (routing capabilities).
MPLS VPRN Service	L3	The routing capabilities in the cloud are provided by the TSP. The utility has to solve the interconnection with the TSP at L3.

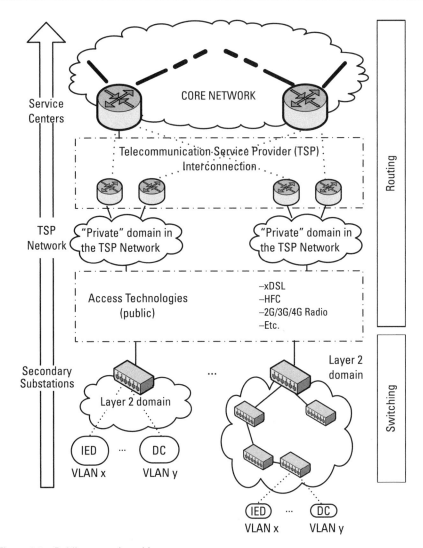

Figure 6.4 Public network architecture.

TSP network and its internal design to maintain the same LANs at the SSs and the same configuration inside the core network to deliver the traffic to the application servers. Because public TSP assets are being used, telecommunication security and reliability has to be considered in detail.

The following is a proposed implementation:

• In public mobile radio connections, the utility connects with an APN [see ETSI TS 100 927 V7.8.0 (2003-09) Digital Cellular Telecommunications System (Phase 2+); Numbering, Addressing and Identification] specific for smart grid services. Every subscriber identity module (SIM) used in a 2G/3G/4G device that carries smart grid traffic has to be registered and activated in this APN. This APN is internally associated by the TSP to a VPN. For wireline solutions provided by the public carrier (e.g., xDSL, HFC), the traffic has to be also added to a VPN (it could be the same as above).

• The traffic carried by the TSP, must be connected to the core network in at least two interconnection points for redundancy purposes. The bandwidth limitation for every interconnection point has to be determined, depending on the amount of traffic aggregated in a single point and the performance requirements. This is an important parameter, because if the expected capacity for the interconnection point is exceeded, congestion will occur and packets will be lost, affecting the TSP provided service.

• At the IP level, the utility and the TSP share IP subnetworks to establish the common network and connect remote devices at the utility interconnection points. This IP connectivity allows traffic routing from the router installed in the SS to the core network. In this shared network the utility has to be coordinated with the different TSPs at the IP level, and any IP address used by the utility shall not be present in the TSP network. To avoid this situation, the recommendation is to use tunneling strategies which allow the utility to maintain a separated IP addressing plan, common for every SS and with no dependencies from the TSP's IP addressing plan. For example, dynamic multipoint VPN (DMVPN) [6] or Layer 2 Tunneling Protocol (L2TP) [7] are possible solutions. Additionally, TSP networks shall provide a translation of the QoS strategies designed for the private network part, in order to have an end-to-end traffic QoS. Last but not least, authentication and encryption techniques (e.g., IPsec) shall be implemented to protect any critical data to be transported through a public network.

• At the core network level, all the traffic needs to be included in the appropriate VPLS or VPRN. The core network administrator will establish policies and relationships among network domains for each traffic type.

6.3.3 Example of a Core Network Design Process

A core network design will eventually define the location of the core routers (MPLS nodes) and their physical interconnection. Obviously, if the utility has an extensive optical fiber plant, premises where plenty of optical fiber cables interconnect, can be chosen for these locations. As an alternative, the use of microwave point-to-point dedicated links, or leased capacity, is also possible. For reliability purposes, it is mandatory to have at least two fully diverse physical paths between every pair of core routers. The implementation of a full-mesh topology is recommended.

The MPLS network aggregates all traffic in the core network; each traffic flow will be constrained to separated particular domains to apply particular policies depending on the assigned class of service. For the purpose of traffic isolation and independency of services (e.g., to guarantee confidentiality), MPLS includes VLLs, VPLSs, and VPRNs services. VPRN service is recommended for the smart grid because network design needs L3 domains interconnection in order to route traffic, as in Figure 6.5. VPRNs are used in this design example. Multiple virtual routers can be defined in the nodes of a MPLS network. These virtual routers share routes among them.

The ACR is the element that connects the access network to the core network. It acts as the default gateway for the hosts in the LAN, and is configured to route the traffic to the next hop in the IP network.

When the ACR is part of the private network, the connectivity between the ACR and the core router (MPLS core) can be implemented with different transport technologies. The recommendation for the L3 architecture is the dynamic routing protocols, avoiding static or default routes; static routing should just be used as an interim implementation due to its low scalability.

There are multiple alternatives for routing protocols (e.g., OSPF, RIP, IS-IS, EIGRP, EGP, BGP [1]). OSPF is usually a good choice but it may have scalability problems (e.g., due to the limitation in the number of neighbors and the amount of microprocessor resources needed [8]); BGP, designed to interconnect autonomous systems in the Internet, is a less hardware demanding alternative. Route announcement among the virtual routers of an MPLS network can be supported by BGP sessions in the control plane, with MP-BGP (Multiprotocol Extensions for BGP, RFC 4760).

Routing protocols only need to act in a single routing domain between the ACR and the MPLS node in the core network (core router). ACRs do not need to share traffic routes among them even if they are connected to the same core router; if needed, traffic between ACRs can be enabled at upper layers. Thus, a single routing domain can be established between every ACR and the corresponding VPRN.

When the ACR is in a public network, it is assumed that the remote site is connected through a tunneling infrastructure (typically in a hub-and-spoke architecture; remote sites are the spokes, and the central part will be the hub). The hub router will be directly connected to a core router of the core network. The routing protocol to announce the private IP subnets aggregated in the ACR to the hub router must be decided. In this case, it is not efficient that the ACR gets to know the routing tables of all the routers, and it can just use a one-way

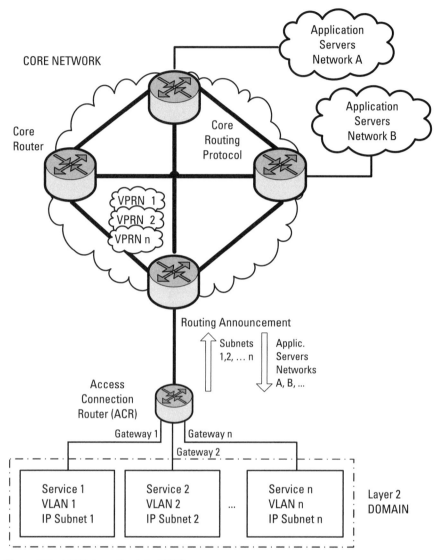

Figure 6.5   LAN integration into MPLS

routing announcement from the ACR to the hub router. Routing Information Protocol (RIP) could be a good choice for this purpose because it works
efficiently in typical hub-and-spoke architectures. In upstream direction, the
ACR should know that all the traffic must be sent to its hub router, defined as
a default route. The connections between the hub router and the core network
can use the same protocol used to connect private ACRs with the MPLS cloud;
the hub router can be considered as a unique private ACR with many routes.

Finally, to deliver the traffic to central application servers, it is common that each service has its own independent server network domain. Thus, the architecture will solve the interconnection points of every service domain following the same philosophy as the internal design of the server network. For instance, if the AMI service has been designed with geographical redundancy (e.g., two main servers in separated premises acting as master-slave or balancing the traffic), the architecture of the core network needs to support such functionality.

For this purpose, the L3 capabilities using routing protocols between the core network and the networks of the application servers can be extended. In order to deliver only the appropriate traffic, it is necessary to decide which the different VPRNs configured to deliver the traffic to the servers are, the interconnection IP networks to establish the next hop, the routing protocol to be used in this interconnection, and the routes to be shared between these routers.

The interconnection of central application servers' network can be challenging depending on how the servers' network has been designed. For example, if the routers of the servers' network are using NAT [as per RFC 3022, "Traditional IP Network Address Translator (Traditional NAT)"], static routes between its own routers or an IP addressing plan not fully compatible with the general architecture, conflicts may arise. Thus, it is important to analyze the possibilities of routing protocols implementation in every segment of the network. An outcome of this analysis could be, for example, to use static routes for a limited number of services, to extend L2 domains (with VPLSs) for others and/or to use dynamic routing for the rest.

6.4   Network Reliability, Resilience, and Scalability

In Chapter 1, network performance, scalability, and reliability were introduced as key factors for network design. The order in which they are mentioned is not casual because it follows a natural flow on the development of networks. The first evident need of a network is to provide the appropriate functional telecommunication services (i.e., performance); once the network is functional, it needs to be scalable for the complete network deployment, and once the services are up and running and can accommodate expected growths, reliability must be guaranteed. However, the sequence above does not mean that the three aspects are not managed in parallel: the analysis and design for performance, scalability, and reliability are iterative, and every decision affects one or more aspects at a time.

Reliability (together with resilience) and scalability aspects are going to be covered in this section, while security is discussed in Section 6.5.

6.4.1   Network Reliability

Because the power grid has obvious, operational reliability requirements, smart grid telecommunication networks have reliability as a major design concern. Reliability affects the whole system; it starts from the devices themselves (e.g., redundant configuration of processors and power supplies, reliable software, strict climatic and environmental requirements) and continues with the network design (from the access part to the application servers and networks, through the core, access, and local networks).

Network nodes and the links between them are collectively called network elements. The probability of failures in network elements affects the reliability of the telecommunication services. The reliability of electronic devices is normally measured by the MTBF, the expectation of the time between failures (see ITU-T Recommendation E.800, "Definitions of Terms Related to Quality of Service"). MTBF is complemented by the MTTR (also defined in ITU-T E.800) referred to how quickly the element can be repaired (or replaced; there is also a MTTF in ITU-T Recommendation E.800 for elements that cannot be repaired). Thus, there is a component in network reliability that is implicit to the electronics used and another component related to the organization when the elements fail and need to be repaired or replaced. Redundancy measures could be needed to improve the implicit reliability features of the network elements through the establishment of multiple node connectivity, multiple physical links (with diverse paths), and backup connections.

Network technologies and protocols must be prepared to implement the mechanisms that automatically use the redundancy options (e.g., the management of multiple redundant physical links). Table 6.4 includes examples of interest for smart grids with such technologies.

Simple architectures may require a limited subset of redundancy schemes. Large networks require the combination of virtually all network redundancy solutions.

Redundancy options must be extended to the TSPs. Independently of the measures that TSPs need to take in their networks and systems, customers may also take their own decisions. As an example, in the context of access networks and 2G/3G/4G radio services, utilities may take the decision to use SIM cards from different TSPs to guarantee that, in the case of total TSP service unavailability, smart grid services may use alternative TSPs.

Critical points for redundancy are the interconnection interfaces with TSPs. Interconnection points, where the TSP is delivering all the traffic to the customer's core network, are critical for service availability. The different sites for the interconnection, the peak capacity of the links, and the protocols in the TSP and utility networks need to be considered.

In the absence of any other criteria, redundancy could be exercised to its highest level, but there are some other aspects that need to be considered.

Table 6.4

Telecommunication Technologies and Redundancy

Technology	Tool	Description
SDH/SONET	Link protection with 1 + 1, or rings	Connection between two nodes may provide a 1 + 1 connection if the links follow different paths, or multiple nodes may build a ring with two paths between every pair of nodes in opposite directions.
Ethernet	Link Aggregation	Two or more links connect a pair of switches (e.g., balancing load). If one fails, part of the traffic still flows.
Ethernet	RSTP, IEEE 802.1D-2004	In an Ethernet network with possible redundant routes, loops are avoided through the use of this protocol that defines the active links and recalculates them if network conditions change.
IP	Routing Protocols (OSPF, IS-IS)	An alternative path can be configured if a node or a link fails in the network. The routing protocol allows the routers to dynamically take the correct decision depending of the status of the network.
MPLS	Fast Reroute (FRR, RFC 4090 "Fast Reroute Extensions to RSVP-TE for LSP Tunnels")	An LSP can be configured as an alternative connection (through a different route) between provider edge routers.

Simple redundancy options must prevail, instead of complex network designs which cannot be easily managed. However, the implementation of redundancy mechanisms needs to be a consequence of a cost-benefit analysis that balances the network areas and the criticality of the services.

6.4.2 Network Resilience

Further to the reliability objective of telecommunication network elements, network resilience becomes the key concept. Resilience starts with reliability [9], but needs to include additional elements to obtain optimal network availability. Resilience is achieved with a combination of prevention, protection, response, and recovery arrangements that include technical and organizational components. Thus, resilience has to do with the ability of an organization to keep the services running in emergency situations to recover them as soon as possible. Resilience has a technical foundation, but it is deeply related to how organizations react and learn from their experience. Resilience needs to focus on the telecommunication system as a whole and on its services end-to-end.

As an example of the multiple dimensions of resilience, we can take the case of network protection and restoration. In a well-designed network, one would expect that in case of failure, the network and its delivered services will be automatically restored. However, as protocols for automatic restoration have their limitations, manual reactive mechanisms must be also ready to be used.

Not all failure scenarios can be predicted and different arrangements need to be in place for such eventualities. Procedures must exist to define how to react to facilitate the efficient and effective restoration of the highest priority services (e.g., the availability of the services of a PS is more critical than those of a SS; remote automation in MV grids is more important than meter readings). However, utilities do not have the same opportunities to drive the restoration of services when the infrastructure is a public commercial one. Although commercial agreements in the form of SLAs should be in place, the ability to drive third-party resources is in general lower than for private networks.

The basic features of resilient telecommunication systems include:

- Architectures with no single points of failure, failover measures (e.g., standby elements) and enough capacity in the network to absorb traffic peaks (traffic, memory, processing capability, and so forth).

- Resilient ancillary services that complement the network (e.g., databases and servers; they are often as important for the network health as the network elements themselves).

- Resilient ancillary infrastructure that supports the network (e.g., housing elements, power supply, batteries).

- Resilient NMSs, as crucial elements for detection and solution of network problems.

- Resilience in the supporting teams operating the network that should have the knowledge and availability to perform their function.

- Resilience in the supporting teams designing the network, which need to constantly review network status and evolution, to adapt designs.

6.4.3   Network Scalability

The telecommunication networks used for the evolution of the smart grid should not inhibit its progress. Current smart grid deployments include applications that have been identified and implemented today; this new applications stimulate new ideas that will bring new applications in an iterative process. It is important to have a well-structured telecommunication architecture that needs minimal changes to evolve.

Not only new services but the constant growth of the number of smart grid premises, and services in existing premises, promote the growth of the telecommunication network. The IP addressing needs to be dimensioned to cope with these unexpected needs. However, network expansion tends to exacerbate the effects of incorrectly designed networks, as design bottlenecks and limits become evident.

Last but not least, an important factor in scalability is the control of the bottlenecks. All network elements and interconnections must be monitored to control the occupancy, processor, and memory usage. These will be early signs of eventual problems that need to be identified and solved before they create any network or service collapse.

The following recommendations may be used in network designs to avoid scalability problems:

- Broadband solutions for multiservice access networks are preferred.
- The design of the general IP addressing plan should take into account that:
  - Every service may grow and eventually need more IP addresses at each delivery point (e.g., substations).
  - New LANs could be needed with the introduction of new applications.
- Scalable protocols should be used. Known limitations of each protocol must be considered in the designs.
- Limitations of network elements for each protocol used must be analyzed. Protocols are often scalable but network nodes could have limitations (processing power, memory needs, and so forth) affecting the performance of the designs.
- Legacy technologies and services are to be avoided, both in the private networks and in TSP services.
- Core network connections must be carefully studied to detect links with capacity growth limits where the network could find a limitation, or where redundancy (real diverse paths) could be needed.
- Interconnection points (with TSPs and with application servers) shall be expansion-capable (e.g., interfaces, connectivity, networks).

6.5 Network Security

Security is a transversal need in a system design and operation, affecting all its parts and processes. The introduction of telecommunications in the electrical grid potentially allows for greater business efficiency; however, wherever ICTs are present, cybersecurity is to be considered.

The smart grid is indeed expanding the connectivity and accessibility of the operation of electrical assets, which can be now reached from different parts of the network without physical contact with the elements. This global connectivity represents also a potential entry point for intrusions, malicious attacks, or other kind of cybersecurity threats.

Deliberate external cyber-attacks are the most obvious security threats. However, the anomalies that may become a security problem have multiple origins such as devices' failures, wrong device configurations, software bugs, user operation errors, and even natural disasters. Security in the smart grid has to be designed holistically to take into account any problem that can affect to the access, confidentiality, integrity, or privacy of all aspects of the electricity business.

Different regulations and standards provide guidelines in order to establish the rules, processes, and protocols as best practices for security in telecommunication systems associated with critical assets.

The security aspects in telecommunication systems for smart grids can be classified around the terminal equipment and in the complete system. Security at application-level protocols has been excluded, as it is not part of the telecommunication network; the IEC 62351 series "Power Systems Management and Associated Information Exchange-Data and Communications Security" can be referenced for this topic.

6.5.1 Terminal Equipment Security

The smart grid is connecting a massive amount of IEDs installed at customer premises (e.g., smart meters) and remote unattended sites (e.g., secondary substations). These electronic devices are connected to a telecommunication network with terminal equipment and get connected to application servers where the system intelligence resides. These telecommunication terminals are a virtual entry point that could be used to access utility systems, hence, the need to consider authentication, authorization, and encryption at terminal equipment level.

Unauthorized access to terminal equipment has to be controlled with the implementation of authentication and authorization. These methods identify who is trying to get access to the device and if the user has a legitimate right to do it; traceability of user actions is also achieved with these methods. The three implications of this are: first, every device in the network must have remote authentication and authorization capabilities; second, it is necessary to have central security systems (e.g., security servers and databases) to manage every remote access to smart grid devices; and third, the telecommunication network must allow these methods.

Further than unauthorized access, transmitted information cannot become available to a third party. For instance, if the transmission medium is the open air (radio), signals could be sniffed by unidentified receivers. Data encryption methods such as 3DES and AES can in general be used. One step further, PKI offers a context for confidentiality in the data exchange. A PKI supports the distribution and identification of public encryption keys, enabling system elements and users to both securely exchange data and verify the identity of the parties.

Physical security cannot be forgotten. Physical access to the network includes both the accessibility to the place where the terminal equipment is installed (e.g., a metering room, an SS, a pole; unattended locations with few access limitations), and the design of the housing of the equipment preventing access to telecommunication interfaces.

6.5.2 System Security

Security is a global concern in telecommunication systems, so security considerations and measures are present in every segment of the smart grid. Every block and area must be protected from the others as potential entry points for cyberattacks. Thus, security policies and specific areas to be protected shall be defined within a network security architecture with different security zones (see Figure 6.6):

- Mission-critical service zone: This zone includes all the electronic devices distributed along the electrical grid that are considered (e.g., SCADA IEDs) and the corresponding centralized or distributed SCADA system.

- Nonmission-critical service zone: It covers the zone related to the applications that are not critical for grid control. This could include, for example, smart metering, LV supervision, and video surveillance in SSs, with their corresponding centralized or distributed systems.

- Private network zone: It includes all the utility-owned core and access networks that have to be managed to provide telecommunications to the service zones above.

- Third-party network zone: This zone consists of the TSPs' networks used. As already mentioned, a tunneling architecture with encryption is recommended (e.g., DMVPN and IPsec).

- Corporate network zone: This is the enterprise area where employees use their corporate nonoperational business applications (e.g., e-mail). Typically, technicians will need access to the operational network to operate and manage the grid from corporate computers in this zone.

- Interconnection zone: This zone includes the interconnection between the different zones or external entities. For example, in this zone, we will find the interconnections between the core network and TSPs or between the core network and the corporate network.

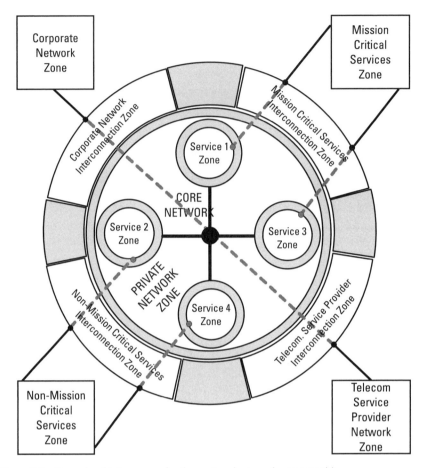

Figure 6.6   Example of telecommunication network zones for smart grids.

Security is especially important in the core network as it must provide the separation between the zones. This separation might be physical, although it would not be cost-effective. MPLS generates this logical separation with VLLs, VPLSs, and VPRNs. For large smart grid deployments with IP as the network protocol, the recommendation is to use MPLS L3 VPNs to provide traffic isolation.

MPLS does not offer data encryption natively. This security aspect has to be solved with other means. Encryption can be provided by link-layer encryption through specific devices. This is a point-to-point approach that is used, for example, in wireless links between the terminals and the router that implements link-layer encryption as part of its MAC layer, or between networking elements using technologies such as Medium Access Control Security (MACsec in IEEE 802.1AE) in the routers [13]. However, these techniques cannot be used with MPLS.

Thus, in the context of MPLS, it is necessary to apply an end-to-end encryption approach, such as the one achieved with IPsec, used in tunnel mode (IPsec encrypts the IP header and the payload) or in transport mode (IPsec only encrypts the IP payload).

The separation between network zones that are part of the smart grid network architecture must have delimiting security elements that protect the resources of one network zone from users of other network zones. This separation is usually implemented with firewalls that control incoming and outgoing network traffic based on a set of rules, and by intrusion detection and prevention systems (IDPSs) that monitor the network traffic looking for anomalous patterns.

References

[1]   Puzmanova, R., *Routing and Switching, Time of Convergence?*, Boston, MA: Addison-Wesley Professional, 2002.

[2]   Szigeti, T., et al., *End-to-End QoS Network Design: Quality of Service for Rich-Media & Cloud Networks*, 2nd ed., Indianapolis, IN: Cisco Press, 2014.

[3]   U.S. Department of Energy, *Communications Requirements of Smart Grid Technologies*, 2010. Accessed March 25, 2016. http://energy.gov/sites/prod/files/gcprod/documents/Smart_Grid_Communications_Requirements_Report_10-05-2010.pdf.

[4]   Goel, S., S. F. Bush, and D. Bakken, (eds.), *IEEE Vision for Smart Grid Communications: 2030 and Beyond*, New York: IEEE, 2013.

[5]   Bragg, A. W., "Which Network Design Tool Is Right for You?" *IT Professional*, Vol. 2, No. 5, 2000, pp. 23–31.

[6]   Cisco Systems, *Dynamic Multipoint IPsec VPNs (Using Multipoint GRE/NHRP to Scale IPsec VPNs)*, 2006. Accessed March 25, 2016. http://www.cisco.com/c/en/us/support/docs/security-vpn/ipsec-negotiation-ike-protocols/41940-dmvpn.pdf.

[7]   Stewart, J. M., *Network Security, Firewalls and VPNs,* 2nd ed., Burlington, MA: Jones & Bartlett Learning, 2014.

[8]   Tiso, J., "Developing an Optimum Design for Layer 3," in *Designing Cisco Network Service Architectures (ARCH): Foundation Learning Guide*, 3rd ed., Indianapolis, IN: Cisco Press, 2011.

[9]   Gorniak, S., et al., "Enabling and Managing End-to-End Resilience," Heraklion, Greece: European Network and Information Security Agency (ENISA), 2011. Accessed March 21, 2016. https://www.enisa.europa.eu/activities/identity-and-trust/library/deliverables/e2eres/at_download/fullReport.

[10] Godfrey, T., Guidebook for Advanced Distribution Automation Communications—2015 Edition: An EPRI Information and Communications Technology Report, Palo Alto, CA: Electric Power Research Institute, 2015. Accessed March 22, 2016. http://www.epri.com/abstracts/Pages/ProductAbstract.aspx?ProductId=000000003002003021.

7

PLC Technologies Supporting the Smart Grid: Use Cases

Power line communications (PLC) is an important telecommunication technology to enable an efficient deployment of the smart grid. Although its penetration is comparatively small in the telecommunication market [where xDSL, hybrid fiber coaxial (HFC), and public mobile cellular radio dominate], this is not the case among utilities because it has been commonly used to provide services when other technologies cannot perform or require significant investments.

PLC has adopted different names and acronyms through its over 100-year history depending on different application domains and performance characteristics: carrier current, power line carrier, PLC, power line telecommunications (PLT), broadband over power line (BPL), and others. The acronyms PLC and BPL will be used as they are the most common ones in the industry.

PLC has always remained amidst the electric and telecommunications worlds. The added complexity of combining state-of-the-art knowledge of both electrical engineering and telecommunication engineering has limited its expansion. A good example is the approach to standardizing PLC taken by various international standards bodies [e.g., International Electrotechnical Commission (IEC), Institute of Electrical and Electronics Engineers (IEEE), International Telecommunication Union (ITU), and European Telecommunications Standards Institute (ETSI)] where relevant initiatives have been limited and not sustained over time.

The relatively small impact of PLC in the broader telecommunications domain can be also considered an advantage because PLC has evolved (particularly in the last decade) to cope with the requirements of the utilities, which can influence its evolution, as opposed to other technologies where utilities are just another user, and a minor one, usually.

There are several considerations around PLC use:

- Broadband PLC versus narrowband PLC: The original PLC networks were designed for carrier frequencies of few tens of kilohertz [1] as the state of the art dictated; progressively, PLC increased its effective bandwidth. One barrier against extending PLC bandwidth was the existence of the AM radio broadcasting band in MF between around 530 and 1,700 kHz. (It has to be noted that in any case PLC, as xDSL, is a conducted transmission.) PLC was never deployed in that frequency range, and this effectively became the dividing range between narrowband and broadband PLC. Broadband PLC has become, with time, bounded only by another broadcast radio band, the FM band starting at 87.5 MHz. Thus, narrowband is the term for any PLC technology below 530 kHz, and broadband PLC exists from around 2 to 87.5 MHz.

- Standardized versus proprietary PLC: As with any other technology, systems exist that are only marketed by a single provider and based on nonpublic specifications. These were for many decades the only available option for PLC users. However, with the advent of smart grid and the tens of millions of PLC units being installed, the interest for int ernationally standardized systems grew and, in the case of narrowband PLC, they currently represent most of the market.

- PLC at different voltage levels: PLC applications in the first half of the twentieth century were analog power line carrier systems for voice and transmission line protection [2]. With time, PLC has been applied to progressively lower-voltage levels, to the point that currently it is massively used in low-voltage (LV) alternating current (ac) systems (e.g., in Europe). Requirements vary for different voltage levels (e.g., from the perspective of coupling signals to the power lines or the conducted power injection limits and the underground and overhead nature of the grids).

PLC is in itself not a technology for the core network, so it needs to be completed with other technologies to provide end-to-end services. PLC has evolved to be a convenient access technology for specific segments of the grid.

7.1 Architecture

The selection of any PLC architecture heavily depends on the physical aspects of the grid.

7.1.1   Grid Constraints

A fundamental aspect of PLC, obviously an advantage for its applicability to smart grids, is that architectures are conditioned by the electrical structure and characteristics of the grid in its different segments. Because ultimately the structure, topologies, types of cables, type of switchgears, voltage levels, deployment practices, and extension of power grids differ in various parts of the world, PLC networks may consider different architectures for the same application depending on the region.

PLC, as an access and local network technology, can be deployed over either medium-voltage (MV) or LV grids. Both grids tend to create tree-like topologies from the PLC perspective:

- Overhead MV grids tend to present bus-like and sometimes meshed topologies. However, from the perspective of cost optimization, seldom do utilities decide to deploy additional PLC nodes for redundancy. In fact, the usual topology [3] basically draws a daisy chain of point-to-point links between nodes that are usually located at relevant locations in the grid (e.g., at transformer poles or close to them). This can be further complicated by some nodes that actually couple to multiple MV lines. In this case, the topology would split at those points and end up forming star topologies with daisy-chain-like branches (which, in turn, could further split and define tree-like topologies).

- Underground MV grids present point-to-point topologies between transformers, with varying degrees of networked topology for redundancy in case of electric faults. Starting from a secondary substation (SS) with a single backbone connection, the PLC topology naturally follows the MV structure in a daisy chain up to a point in which some SS has several outbound MV lines with deployed PLC links, so the final structure would end up resembling a tree structure. Some experiences exist with redundant ring topologies.

- LV grids, either underground or overhead, usually present complex topologies with multiple branches and subbranches in many hierarchy levels. Some branches connect to each other, forming ring-like topologies. Still, the LV grid is usually originated from a single source (i.e., the secondary substation or SS) so a PLC master node is usually placed there, and then PLC branches are defined over the LV lines, in subsequent hierarchy levels that split as far as the final end-points. These end-points are final customers.

Tree-like logical topologies allow for signal repetition/relaying and increased reach/coverage both in MV and LV scenarios; they are a fundamental feature of functional PLC networks. Depending on the type of grid (i.e., overhead or underground), the possibilities to connect devices to perform this repetition are different. If the repetition is to happen on the overhead grid, direct access to the cable for coupling and installation is relatively easy. If the repetition is needed in an underground grid, the devices can only be placed at points where the grid can be accessed (e.g., fuse boxes, street cabinets, meter rooms).

A major effect to consider is the propagation characteristics of the power lines in each of the scenarios (MV or LV). Cable design and disposition (e.g., overhead lines show generally better propagation characteristics than underground ones) and the proximity of noise sources (e.g., disturbance from uncontrolled noisy end-points presents almost exclusively on LV grids) need to be considered to optimally understand PLC signal propagation and maximize signal-to-noise ratio (SNR).

The electricity grid is a truly interconnected network at 50 Hz or 60 Hz: electricity flows from the MV to the LV segment through transformers. However, PLC systems use higher-frequency signals so connectivity behaves differently; this will affect system planning (e.g., frequency bands, coexistence, and so forth). Legacy UNB PLC systems below 10 kHz could usually inject signals at PS feeders and have them traverse transformers, so that the attenuated signal could still be detected at LV end-points. PLC signals in the range of tens or hundreds of kilohertz can be sometimes partially transmitted from the MV segment to the LV segment and vice versa [4]. A signal of some megahertz may normally not cross a MV/LV transformer. The frequency behavior of the signal transfer from MV to LV depends on the transformers characteristics, impedances, and other installation factors; it is in general unpredictable if PLC signals are going to be naturally coupled from the primary to the secondary winding of the transformers.

7.1.2 PLC Architecture Domains

The combination of PLC technologies for each segment of the grid has to be decided. The final configuration will be based on technical factors such as the bandwidth needed for the services, the existing electric grid infrastructure, the existence of adequate technology and vendors, and the preferences of the utility. No single architecture fits all utilities.

Usually MV and LV PLC networks are considered physically separate (for frequencies above tens of kilohertz). The same generally applies for LV networks supplied by different MV/LV transformers. As a reference of a common PLC architectural implementation, a certain PLC system is implemented in the LV grid, and a different one is implemented in the MV grid to extend the telecommunication network to central systems; the two PLC domains, the one in the LV grid and the other in the MV grid, are independent.

Table 7.1 shows the most usual combinations of PLC technologies for MV and LV domains, along with specific applications.

From a utility perspective, the first step in considering deployment of private PLC solutions as part of the overall smart grid telecommunication network would be to understand where PLC networks should be cost-effectively deployed. Several situations are possible and actually being considered and deployed by utilities worldwide. Considering a conservative approach with little concern for investment constraints, deployment of BPL both in MV and LV segments would be advised for highest performance. If smart metering is expected to be the main driver for the LV grid in the foreseeable future, narrowband (NB) PLC in the LV part may become the most sensible choice from an investment perspective, possibly combined with BPL or even NB PLC for the MV grid. A utility may think of combining approaches with regard to BPL/NB PLC usage in MV/LV grids for different parts of its service area. For specific scenarios in certain grids, UNB PLC remains a valid alternative [5–7]. SSs are network nodes for these deployments (Figure 7.1).

7.2 Design Guidelines

Design aspects for the four possible PLC application scenarios of Table 7.1 will be discussed focusing on the use of BPL over MV grids and NB PLC over LV grids as the preferred combination.

7.2.1 Broadband PLC Networks

BPL networks use technologies that transmit and receive signals using frequencies above around 2 MHz and up to several tens of megahertz. Although BPL can be applied to LV in many scenarios, this section starts discussing the use of

Table 7.1
Combination of PLC Technologies and Grid Segments

PLC	Low-Voltage Grid	Medium-Voltage Grid
Narrowband	Smart metering and smart grid applications over LV grids: IEC 61334-5 series, ITU-T G.9904 (PRIME), ITU-T G.9903 (G3-PLC), IEEE 1901.2	Smart metering applications, low data rate DA: IEC 61334-5-4, ITU-T G.9904 (PRIME), ITU-T G.9903 (G3-PLC), IEEE 1901.2
Broadband	Internet service, smart grid applications in some scenarios: IEEE 1901, ITU-T G.hn, OPERA technology, ISO/IEC 12139-1	Applied for smart grid backhauling applications: OPERA technology

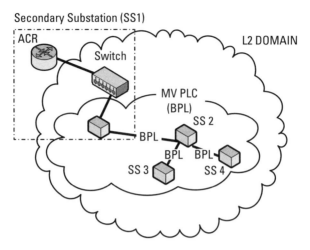

Figure 7.1   PLC architecture example.

BPL on MV: this part of the network is more controlled and predictable than LV.

7.2.1.1   Medium-Voltage BPL

The technology described in this section is based on the OPERA-defined specification (see Chapter 5), currently the most proven and mature technology for MV BPL.

Signal Coupling

Coupling of PLC signals into MV lines is a unique factor to consider in the deployment of BPL. The purpose of coupling is to inject BPL signals into the MV grid to be propagated to remote peers over it.

MV BPL couplers are defined as the elements that allow BPL signals to be superimposed onto and/or extracted from a power waveform while also preventing the power waveform from entering the transceiver. BPL signals employ high frequencies so the values of relevant coupling elements (capacitors, inductors, and so forth) have to be designed accordingly. Two possible coupling methods are usually considered: inductive and capacitive.

Inductive coupling is flexible and easy to install; it is a low-cost solution that makes it attractive for some applications. Inductive couplers are electrically insulated from the power cables: usually toroidal magnetic core halves are clamped around the MV cable. They usually are clamped around a single MV phase cable, grounded on their ground terminal and connected to a PLC device through a 50-ohm coaxial cable connector. However, inductive couplers (see Figure 7.2) are not widely used for MV BPL as circuit behavior may vary with

(a) MV Capacitive Coupler for   (b) MV Capacitive Coupler
Air Insulated Switchgears   for SF6 Insulated Switchgears

Figure 7.2   MV BPL couplers.

line impedance to the point of possibly stop functioning when any of the line breakers in the circuit is open.

Capacitive coupling (see Figure 7.2) is used in most smart grid applications, due to the need to have reliability independently of the state of breakers and switches. Capacitive couplers (see Figure 7.3) can be installed in a wide range of scenarios and allow for a very effective coupling of high-frequency (HF) broadband waveforms. Coupling capacitors are designed with their capacitance ranges between 350 and 2,000 pF (the higher, the more expensive and the better performance). MV BPL capacitive couplers should be tested against IEC 60358:1990 and IEC 60481:1974. The larger the capacitance, the more complex and expensive a coupler is, and usually the better its performance in the low part of the BPL frequency band. Capacitive couplers normally inject signals directly between one phase and ground. This implies a relatively simple

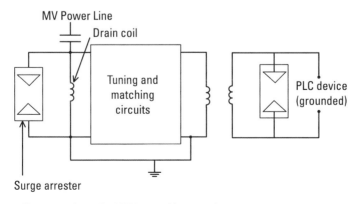

Figure 7.3   Elements of a typical MV capacitive coupler.

installation process in the case of older masonry and open-air (metal-clad and metal-enclosed) switchgear.

For modern MV gas insulated switchgear (GIS), specific MV BPL couplers exist in the market. The usual method for installing the coupler is directly into the standardized separable connector (e.g., according to EN 50181:2010 in Europe), which fixes the bushing to the cable.

MV BPL couplers are normally required to interface to the BPL device with a 50-ohm coaxial cable connector. Additionally, they need to be constructed with a robust design that allows for long periods of unmaintained operation (up to 25 years) and for compliance with safety requirements.

Because of significant differences between MV distribution grids, it is not uncommon for MV BPL coupling solutions to need some customizing to the type of MV assets that a specific utility owns, which usually include underground and overhead grids with associated different switchgear. IEEE 1675-2008 specifically addresses the U.S. market.

Considerations on Overhead and Underground Grids

BPL does not behave the same in overhead and underground MV links. From the signal propagation perspective, overhead MV grids are preferred to underground. However, underground MV is actually favored in this use case, to the point of MV BPL being deployed exclusively over underground grids. The reasons are that:

- Underground lines usually have a single, point-to-point path between two SSs. No stubs or derivations or switching elements exist in the middle. This is an advantage with respect to overhead lines in which normally derivations (taps) exist in the line path between any two transformers.

- Distances of underground lines, normally found in urban and semi-urban areas, are shorter than for overhead lines. This generally allows for controlled attenuation levels in a BPL link between two SSs and makes propagation concerns irrelevant.

- Underground lines are more immune to potential interference from radio services than overhead ones.

- Underground lines are terminated, in the case of European networks, by SSs, which are usually well-protected locations with normalized switchgear and direct access to the MV lines. This is contrary to the case of overhead MV grids where working at heights (to install MV BPL couplers) becomes a safety issue with cost implications.

- The return on investment is maximized in underground lines as they usually correspond to urban SSs where higher meter densities are found.

Network Planning Criteria

Contrary to a common belief, the main challenge for MV BPL is a proper network planning methodology that ensures performance within predefined limits. Planning rules and guidelines need to be set, considering the specific frequency division multiple access (FDMA) or time division multiple access (TDMA) technologies used.

Starting with the different possible approaches to manage the access of multiple nodes to a shared and difficult medium, BPL has historically adopted different solutions for physical (PHY) and medium access control (MAC) layers. The two standard options specified in IEEE 1901-2010 (i.e., one based in Homeplug [8] and the other in HD-PLC [9]) have not seen, up to now, wide adoption in the MV BPL space. The adaptation of ITU-T Recommendations G.9960 to G.9964 [10] to MV BPL networks is yet to be seen, if and when compliant devices become available. The industry solution that has been tested and demonstrated through large scale deployments in field since 2008 is the Open PLC European Research Alliance (OPERA) technology [11] (see Chapter 5).

The general concept of OPERA technology for MV grid deployment is through implementation of limited-size TDMA domains that coexist one with another through the use of an FDMA approach that guarantees noninterference between close TDMA domains. The number of FDMA bands should be the minimum necessary (ideally, just two). This case study assumes that the employed BPL technology is based on two distinct frequency modes such that a BPL cluster will be defined as the set of BPL devices that share a single TDMA domain while transmitting and receiving in the same frequency mode. Obviously, two clusters using the same frequency mode shall be far enough from each other so as not to interfere. The problem of network planning becomes similar to one of radio cellular network planning.

The technical choice to optimize the use of available frequencies is an important consideration for MV BPL planning. A careful selection of the injected power levels (which have to comply with regulation) and frequency modes and a knowledge of the effects of MV cable characteristics (conducting materials, configuration, additional grid elements) on BPL propagation provide useful confidence levels that actually validate the choice of a systematic planning methodology. Based on theoretical analysis and extensive deployment experience [11], the recommended definition of the two frequency modes is such that Mode 1 will use the 2–7 MHz range and Mode 2 will use the 8–18 MHz range. Despite the apparent asymmetry, both result in similar bandwidths in real deployments due to the channel characteristics (see Figure 7.4).

In order to guarantee performance and latency requirements for the typical smart grid applications in MV and LV grids, a total of eight independent criteria are used for the planning of MV BPL networks:

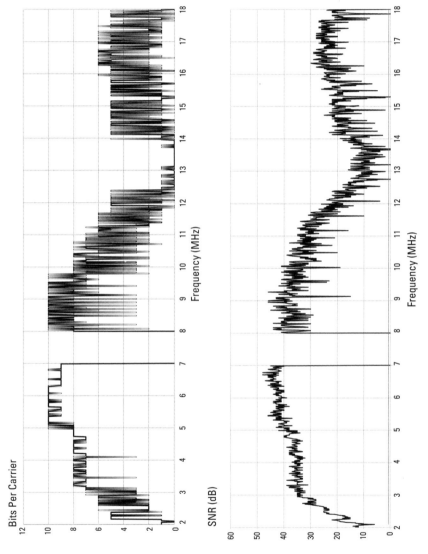

Figure 7.4   Sample MV BPL channel response in the 2–30-MHz range.

- Maximum number of SSs per cluster (S_{MAX}): In a shared TDMA domain, the number of nodes must be limited to control low throughputs and high latencies.

- Maximum number of hops between master and slave (H_{MAX}): Every cluster is defined by one single master node managing the beacon and assigning time slots for transmission. The remaining nodes in the cluster are called slaves. In order to be able to guarantee latency in a shared TDMA domain with packets repetition, the maximum number of hops between a master and any of its slaves shall be established as H_{MAX}.

- Maximum number of consecutive SSs (S_{CON}): Any cluster shall comprise a maximum of S_{CON} consecutive SSs in any chosen path. It is added to further limit the performance decrease implied by multiple long branches.

- Maximum distance: It is the distance that the BPL links can reach. Conservative approaches are suggested; the values recommended here are based on extensive experience with typical underground European MV grids. "New" MV cables refer to modern, polyethylene (PE)-insulated [e.g, cross-linked polyethylene (XLPE) or ethylene propylene rubber (EPR)] or polyvinyl chloride (PVC)-insulated single-core cables, and "old" cables are usually the classic paper-insulated, lead-covered (PILC), oil-impregnated, three-core cables:

    - Mode 1 can be used for new MV cables in links up to M_{MAX1} meters. For Mode 1, 1m of old cable segment counts as m_1 meters of new cable, as new cable propagates signal better.
    - Mode 2 can be used for new MV cables in links up to M_{MAX2} meters. For Mode 2, 1m of old cable segment counts as m_2 meters of new cable.
    - Duct sharing can create interference in cables running in parallel. Two BPL links in different clusters using the same mode cannot share the same duct.

- Minimum guard distance: It guarantees that there is enough physical distance (through MV cables) between two clusters using the same mode to avoid interference that could be induced through switchgear and propagate through MV cables. The criterion is designed to be rather conservative. Note that old cables allow for shorter minimum guard distances, as BPL signal propagation is worse:

    - The guard distance (for new cables) between two different clusters in Mode 1 shall be more than G_1 meters. For Mode 1, 1m of old cable segment counts as g_1 meters of new cable.

- The guard distance (for new cables) between two different clusters in Mode 2 shall be more than G_2 meters. For Mode 2, 1m of old cable segment counts as g_2 meters of new cable.
- Duct sharing: For the calculation of guard distances above, the distance of parallel cable runs (possible interference in duct sharing) shall be subtracted from the total guard distance.

- No customer-owned SSs: It is common to find, in underground MV grids, some SSs that are not owned by the utility, but rather by a large customer supplied in MV. To avoid operation and maintenance problems, this criterion ensures these SSs will not be part of any designed BPL cluster.

- Maximum number of nonserviceable SSs (C_{MAX}) per cluster: If an SS is of no interest for the smart grid, it is defined as nonserviceable. These SSs do not need any telecommunication service, and the only reason why they would become part of a cluster is because MV BPL usually needs to be deployed at every sequential SS in a MV line path to avoid the attenuation of the switchgear in their path from one MV line (BPL link) to another. Any single cluster shall comprise a maximum of C_{MAX} nonserviceable SSs to maximize useful investment.

- Minimum number of serviceable SSs (S_{MIN}) per cluster: The motivation of this criterion is economic, as there must be a minimum size to the BPL cluster that shares a single backbone connection at the master to maximize its economic efficiency compared to other technology options.

A practical way to massively execute network planning with MV BPL is through automated tools; it would not be possible otherwise to define optimal assignment of BPL resources to smart grid deployments in the order of tens (or even hundreds) of thousands of SSs. For such an automated tool, some merit areas can be defined to qualify the various BPL plannings:

- Expected performance: The main parameter to predict the performance of the cluster is through its links (distance and cable type). For every planned BPL link, the equivalent link distance (in terms of new cable and considering either m_1 or m_2 for the old cable segments) shall be calculated. The shortest the total equivalent distance, the better.

- Number of serviceable SSs and necessary clusters: It measures how many serviceable SSs are connected with MV BPL and how many clusters are needed to do so. For a constant number of serviceable SSs connected with MV BPL, the higher number of clusters, the better.

- Expected backbone availability: Each planned cluster will need an SS with an access connection router (ACR) connected to the core network (see Chapter 6). The planning of clusters that will have real possibilities of being delivered a performing connectivity will be favored.
- Total number of smart meters covered with BPL: The largest total number of LV customers, the better.

7.2.1.2 Low-Voltage BPL

Utilities have been using LV BPL since the 1990s [12] for two main applications:

- Internet access in competition with wireline technologies such as xDSL or HFC [13, 14]: This service did not meet wide commercial success; a side application started taking off commercially in the 2000s for in-home communications between consumer electronics devices such as personal computers (PCs), DVRs, and television (TV) sets. BPL has the ability to overcome Wi-Fi technologies' coverage problems [15, 16], and more than 100 million chipsets have been shipped to the consumer electronics industry [17].
- Smart grid services, mainly smart metering.

Use of BPL in LV grids is conceptually similar to MV, with some specific features:

- LV grids are less controlled by utilities. It is harder to provide a network planning, based on simple, common criteria, which performs as expected once deployed.
- LV grid topologies are more complex (European grids), with higher number of branches and subbranches extending radially. This implies signal reflections at high frequencies. This can be compensated by installing repeaters at link boxes and other accessible locations.
- Utility budgets for LV grid maintenance are lower than for MV (usually just reactive maintenance is done at LV); in some areas LV cables may be degraded affecting BPL signals.
- It is not possible to have a control of loads directly connected to LV grids (as opposed to the MV case). This is a potential source of problems due to unknown terminal impedances and noise levels.

In conclusion, guidelines given for MV BPL cannot be directly extrapolated for LV BPL, although from the perspective of network design and deployment, there are obvious similarities.

For utility deployment of LV BPL in the smart grid context, the definition of threshold data rates is similar to the MV case. Instead of SSs, LV BPL nodes would be installed at meter rooms (e.g., at the basement of buildings) or inside individual meters, all of them accessing the LV master at the SS through LV repeaters.

7.2.2 Narrowband PLC Networks

Narrowband (NB) PLC is the oldest, most mature PLC technology and the first one to have practical and commercial use. It has been employed by utilities in HV for almost 100 years for voice telecommunications and critical operation purposes. Recent interest is mostly concentrated on LV applications, although some scenarios may find the application of NB PLC on MV interesting.

7.2.2.1 Medium-Voltage NB PLC

NB PLC started being applied to MV with the advent of joint efforts among utilities to standardize distribution automation (DA) in the late 1980s. The concept of distribution line carrier (DLC), as opposed to power line carrier, which was assumed to apply exclusively to HV and transmission lines, was applied to the whole 1990s' IEC 61334 series that is behind the original concept of DLMS/COSEM for smart metering. IEC 61334 series explicitly mention and define usage of NB PLC over MV grids (e.g., IEC 61334-5-2 and IEC 61334-5-4).

None of the two technologies achieved any real success in its applicability to MV as BPL was already being proposed for LV, and even most simple automation and control applications were already expected to require throughputs at least in the tens of kilobits per second.

The scenarios in which MV NB PLC has raised some interest are usually U.S.-type grids in which the distribution grid is mostly MV, so LV segments are limited in extension and number of meters. Several LV segments, each supplied by a single transformer, would be part of the same NB PLC network; the location of the backbone is usually somewhere in the MV grid, and as different LV segments are to be part of the same network, it becomes mandatory that transmitted NB PLC signals cross through transformers, with the help of simple passive devices.

In the specific case of U.S. utilities, it has to be mentioned that FCC Part 15 regulates an advantageous situation for NB PLC: conducted limits for carrier current systems which operate below 30 MHz only exist for the 535–1,705-

kHz range (see 47 C.F.R. §15.107 (c)). There are no established limits for other frequency ranges (radiated emission limits do exist, though).

The direct coupling of NB PLC signals on the MV grid draws from the experience of HV coupling and several products are available in the market (see Figure 7.5). Special care has to be taken with grid elements such as breakers, which may become open and impede signal propagation through the MV lines.

7.2.2.2 Low-Voltage NB PLC

The typical use of NB PLC in the smart grid context is for smart metering applications over LV grids [18]. NB PLC nowadays provides tens, if not hundreds, of millions of smart meters with the ability to remote measured-value readings and also the ability of remotely connect and disconnect internal breakers. Remote firmware upgrade of the smart meters themselves also becomes a possibility with NB PLC, so the smart meter becomes a dynamic and upgradeable node which will perform additional functions in the future.

Information on real deployments and actual results of utility-wide, large-scale NB PLC networks (both standard and proprietary) is not as publicly available as it could be expected. With regard to research in this field, the literature tends to focus more on theoretical aspects of the underlying techniques (channel and noise models, modulation and coding, MAC strategies and protocols) than in real system field analysis.

Figure 7.5   MV NB PLC capacitive coupler.

Technology Description for Grid Topologies

To discuss the massive deployment of LV NB PLC for smart grid applications, we have to previously consider the specific features which make these networks unique.

LV NB PLC channels have been extensively studied [19] but it is virtually impossible to give a simple formulation of a channel model that also includes stochastic noise, which can really be applied to any kind of LV grid. In the absence of a predictive, easily applicable and useful model, LV NB PLC field deployments are advised to lean on conservative choices: low-order modulations with simple yet powerful channel coding schemes. One example with several instances of massive deployments is Powerline for Intelligent Metering Evolution (PRIME), which in the case of European Committee for Electrotechnical Standardization [Comité Européen de Normalisation Electrotechnique (CEN-ELEC)] A-band (3–95 kHz according to EN 50065-1:2011) tends to configure transmission on field at 20 kbps using frequency-domain differential binary phase shift keying (BPSK) and a ½ rate convolutional code. This kind of modulation and coding scheme (MCS) allows for a certain SNR margin on average LV channels so that, similar to radio, several decibels of instantaneous fading can be endured.

The selection of modulation and coding schemes is an important matter to provide the optimum performance in a hostile channel [20] with limited frequency range and allowed power levels, while in the context of cost-effective solutions. The main challenge of LV NB PLC is to specify robust signaling methods which allow for the longest propagation distances and reception under typical LV noise conditions, without unacceptably compromising performance (i.e., throughput).

Available multicarrier modulation (MCM) schemes, almost exclusively based on one or other form of orthogonal frequency division multiplexing (OFDM), generally specify adaptive modulation options. This takes advantage of the state of the art of electronics to try to maximize throughput for any time and location. However, this is not necessarily a fundamental benefit in the case of NB PLC, as constantly changing grid conditions pose a real challenge to any algorithm that tries to establish instantaneous channel conditions and then selects the best of a possible array of MCSs. Other than the algorithm needing some kind of information overhead, which consumes either time or frequency resources, it has been seen that LV noise patterns appear and disappear in a matter of milliseconds, and impedances are in a constant state of change.

To overcome the limitations of LV NB PLC channels, some higher-layer mechanisms are recommended:

- Segmentation and reassembly (SAR) mechanisms: Short packets have a much higher survival rate in LV NB PLC channels (not least because of the inherent short noise burst generated with each zero crossing of the power waveform, at 50 or 60 Hz) [21]. Thus, it is beneficial to have a way to segment long upper-layer packets in a way that they become shorter-length packets (less prone to be affected by noise) in the channel and can be reassembled at the receiver. This approach will slightly increase latency and overhead, but ensures robust system availability (stability) and success rates in different conditions.

- Automatic repeat request (ARQ) as part of the MAC layer: Reference [21] showed the benefit of applying ARQ in this kind of scenario, further to robust MCSs. There is an inherent overhead in every MAC layer packet needed to implement it.

From a system-level perspective, some important aspects need to be considered:

- In the overall analysis of performance, it has to be taken into account that some references [22] reckon that the actual system throughputs will be mostly limited, above all, by the layers above PHY. In a typical DLMS/COSEM smart metering system, the overhead introduced by the polling process to collect smart meter readings makes the net data rate orders of magnitude lower than the raw PHY rate. The efforts to improve performance need to be focused on the upper layers rather than at PHY layer, where the effect will not be visible.

- LV NB PLC networks are configured with some kind of master node (e.g., BN as defined by [23]; coordinator in other systems). This network element typically injects signals at the point with best access to all relevant LV segments (i.e., the MV/LV transformer in the SS, on the secondary winding of the transformers). Because the end-points are normally smart meters (or telecommunication modules that concentrate several of them), their location is also fixed and not to be changed. Virtually all NB PLC modern solutions have the ability to have network nodes retransmitting packets to other nodes downstream in the network. For example, PRIME defines that every service node (SN) (the PLC device in each smart meter) in the network may transition [after a decision by the base node (BN)] to a switch status, in which the MAC layer will repeat packets addressed to distant nodes. Other approaches [24] implement repetition features from a L3 perspective.

Signal Coupling

PLC signals are expected to propagate for long distances in the LV segment. One single SS may feed just one isolated LV segment in a bus topology, or sometimes multiple LV segments may be interconnected together and then, in turn, to various SSs. In terms of topology, European LV grids are larger and more complex than MV ones, so it is not uncommon to find correct reception of PLC signals transmitted from other SSs. At the same time, special cases may have to be considered when, for example, two or more MV/LV transformers share a single SS location and may or may not supply overlapping LV segments.

In a typical SS there may be more than one LV panel (also called fuse board or LV bus bar), associated to one or more transformers (see Figure 7.6). Multitransformer scenarios are typical of urban areas with high customer density. LV panel injection in multitransformer environments has been addressed in [25]. Connectivity among multiple transformers in an SS cannot be sometimes easily determined, and, in any case, NB PLC signals being injected into one LV panel and then crossing two transformers and coupling to another LV panel work unreliably at best. A possible solution is an extension of the concept of repetition, such that one node acting as a master injects signal into one LV panel, and all other LV panels in the SS are directly injected by respective auxiliary nodes, which directly connect to the master node (e.g., through Ethernet or other means).

The strategy of auxiliary nodes, which has successfully been implemented for massive smart metering rollouts, does not rule out however the possible use of multiple master nodes even in SSs with multiple LV panels and TDM networks. The reason is that for sufficient levels of isolation between LV panels, the actual interference between unsynchronized TDM PLC networks using the same frequencies is limited and there is an advantage in using multiple masters simultaneously taking care of different parts of the LV grid, in terms of latency and meter reading collection times reduction.

With regard to the PLC signal injection of (master or auxiliary) nodes into SS LV panels, European grids usually show standard LV panels with several

Figure 7.6   LV panel at European secondary substations.

three-phase and four-wire distribution cables (i.e., LV feeders or "distributors"). LV panels connect to the secondary winding of transformers on one side and are made of a metallic frame that supports four bars (three phases and one neutral) where LV feeders connect. Typical PLC injection is in differential mode between phase and neutral or phase and phase.

Single-phase signal coupling is preferred to three-phase coupling [26], due to signal injection loss and worst-case scenario in reception. PLC signals get coupled capacitively and/or inductively to other phases in cables running in parallel for long distances.

7.3 PLC Technology Use Cases for the Smart Grid

There are multiple smart grid initiatives of utilities around the world that employ PLC both in broadband and narrowband. This section focuses on two real use cases that can directly be used as a reference for typical European grids. Consequently with the guidelines given above, BPL is used for MV scenarios and NB PLC for LV smart metering.

7.3.1 Broadband PLC Telecommunications over Medium Voltage as a Multiservice Backbone

The strategic decision to use BPL on MV for smart grid may be seen as natural for utilities: the ownership of the transmission medium (i.e., power lines) is a huge advantage in itself. Examples exist that have taken it to practice [27], as the technology is mature.

This section covers a use case with tens of thousands of MV BPL links providing smart grid services, showing a high degree of technology resilience, process maturity and operational efficiency. Methodologically, the architecture principles and design criteria already discussed were followed in an iterative process such that other utilities can follow a similar approach, adapting criteria to their specific needs and context.

7.3.1.1 Design Criteria Validation

As already explained, BPL network design is supported by design criteria that have to be adapted to the specific features of the MV network (topology, type of cable, type of switchgear, age of assets, and so forth) and to the characteristics of the BPL technology to deploy (frequency ranges, data rates, scalability, and so forth). The actual values to be used for large-scale programs must be verified with real experience so the initially calculated values are seen to fit with the real grid, achieving expected performances.

In this use case, the validation was made through a specific pilot with SSs in one representative area, where BPL devices and couplers were installed.

Extensive performance measurements were done. The initial boundary conditions for network planning were:

- Only underground MV cables were considered. The topology is less complex for underground grids, and network design is simpler.

- A minimum of simultaneous and bidirectional 100 kbps should be obtained at every SS. This is a data rate coherent with the assumptions in Chapter 9 for advanced metering infrastructure (AMI) and DA applications.

- Maximum number of devices (MAC addresses) to exist in a single L2 domain (this includes BPL devices and all other MAC addresses in their respective LANs) would be 1,024. This is a non-BPL related limit, with origin in one chipset implementation.

- The raw PHY data rates individually obtained in any link, either downstream or upstream, should always be at least 10 Mbps. This is because it is observed that the deployed BPL technology may become unstable if PHY rates were below, and it is also related with the 100 kbps as an application layer requirement.

- Network planning considers two frequency modes, Mode 1 (2–7 MHz) and Mode 2 (8–18 MHz).

- Network planning criteria (see Section 7.2.1.1) are in Table 7.2.

The tested network topology is shown in Figure 7.7, where the SSs, the MV lines between them, and the BPL clusters are shown.

Inside this cluster, distances between SSs varied from 100m up to 500m, and PHY rates were measured in all cases to satisfaction: in Mode 1 links, the average was 30 Mbps (bidirectional), whereas Mode 2 presented 58 Mbps average (higher standard deviation with Mode 2, due to the difference in the PHY rates of the shorter links with respect to the longest ones). Table 7.3 shows the derived attenuation values at different frequencies.

It was also observed that power lines with newer cables, which had fewer repairs done and a single type of cable in the link between two SSs, performed

Table 7.2
BPL Network Planning Parameters

Parameter	Value	Comment
S_{MAX}	20	Maximum number of SSs in a single BPL cluster
H_{MAX}	9	Maximum number of consecutive hops to a master
S_{CON}	15	Maximum number of consecutive SSs in any given path
C_{MAX}	1	Maximum number of non-serviceable SSs per cluster
S_{MIN}	4	Minimum number of serviceable SSs per cluster

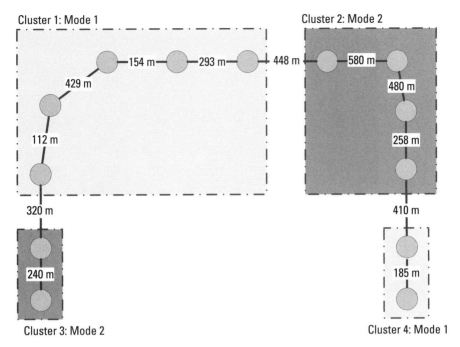

Figure 7.7 MV BPL use case pilot to validate criteria.

Table 7.3
MV BPL Attenuation at Different Frequencies

Frequency (MHz)	Calculated Attenuation of 100m of MV cable (dB)
2	1.5
7	5.5
8	6.3
18	15.0

better than older cables with one or more joints. However, getting a detailed knowledge of the exact condition of underground MV lines is not easy for most utilities, so it was important to just define a conservative criterion and apply it for all cables based on a simple division between new and old cables. Based on that, the values given to parameters were:

- For Frequency Mode 1, $M_{MAX1} = 1,000$m and $m_1 = 2$m. This means the maximum distance Frequency Mode 1 BPL will be assumed to reach in an old MV cable is 500m. $G_1 = 2,000$m and $g_1 = 1.5$m.

- For Frequency Mode 2, M_{MAX2} = 700m and m_2 = 2m. This means the maximum distance Frequency Mode 2 BPL will be assumed to reach in an old MV cable is 350m. G_2 = 1,500m and g_2 = 1.5m.

The final step was to check that the application throughput in all SSs in the pilot was above 100 kbps. For this, commercial computers were installed at every BPL-connected SS. The computers ran software that generated and measured traffic in different conditions [e.g., half or full duplex, User Datagram Protocol (UDP) or Transmission Control Protocol (TCP), uplink or downlink] and they were all synchronized so as to simultaneously load the BPL network with traffic from all the computers. The results were satisfactory so that all SSs were able to simultaneously provide at least 200 kbps.

7.3.1.2   Network Planning

Once network planning criteria for MV BPL in underground lines, along with figures of merit to compare network designs with each other, were defined, BPL plans for massive deployment were made.

One-line diagrams are needed in underground MV grids to represent the MV logical connectivity among SSs and the equivalent link distance assigned to each cable run between SSs. Then clusters of BPL-connected SSs must be defined.

The decision on the cluster formation can start from the SSs where backbone connectivity is more easily available. From there, clusters should try to cover as many SSs as possible while expanding towards the edges of the considered area. Figure 7.8 represents a BPL design over a certain area.

7.3.1.3   Results Obtained in Large-Scale Deployment

Some 14,000 BPL devices have been installed in the field in the example use case under study, with an increment of 150 every week.

Statistical results completely validate the design criteria. Figure 7.9 shows results of the 14,000 MV BPL links deployed. The statistical data are directly reported by the BPL network management system and they show instantaneous BPL link measurements at the time the information is requested. Over 40% of the links performed between 30 and 50 Mbps. The latency of the BPL links, as a rule of thumb, is of some 10 to 20 ms in each BPL hop.

Any nonperforming BPL link is checked on the field. A short list of common problems and solutions is shown in Table 7.4.

7.3.1.4   High Availability in Medium Voltage Broadband over Power Line

The BPL cluster design proposed in the use case has a single point of failure at the master location (i.e., one of the SSs of the cluster) where the backbone connection exists. Reference [28] provided a rigorous study on this issue, analyzing

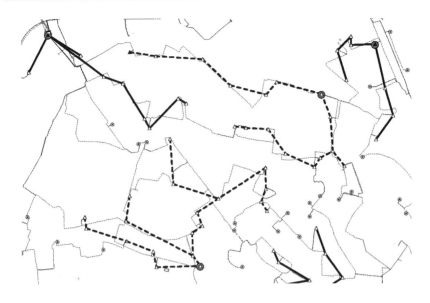

———— Mode 1 BPL Cluster

– – – Mode 2 BPL Cluster

△ Secondary Substations in a Cluster

⊚ Secondary Substations with BPL Master Device and Backbone Connection

⊙ Secondary Substations with public radio connectivity

⸺ MV Feeders

Figure 7.8 Example of MV BPL network plan.

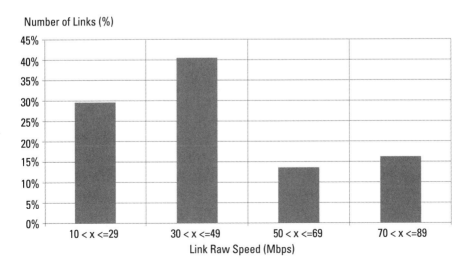

Figure 7.9 BPL links PHY performance (uplink and downlink).

Table 7.4
Typical Problems in MV BPL Links

Category	Symptoms	Solution
Impedance mismatch	Low received power levels (OFDM carriers), asymmetrical (both ends) received power levels (OFDM carriers)	Cabling (connectors) problem solving, coupling unit installation improvement (cable distance shortening)
Saturation	Short MV cables distance, low performance (throughput), reception automatic gain control = 0	Attenuation by software, attenuation using external hardware
High Attenuation	High MV cables distance, low performance (throughput), low SNR, reception automatic gain control = maximum	Cabling and coupling units connection improvement, MV cable phase change
Noise	Low performance (throughput), low SNR, reception automatic gain control = maximum, abnormal BPL received signal spectrum	Cabling and coupling units connection improvement, noise filtering (e.g., mains)

data from field operations. There are three types of failure cases considered (see Figure 7.10), derived from real operational data:

- Backbone fault: Comprises 33 % of the total amount of MV BPL network fault events.
- BPL master fault: It is representative of 6% of network fault events.
- BPL link fault (cluster split): It represents 61% of network failures.

Based on the analysis of the failures, it is possible to set strategies that allow BPL clusters to be resilient to network faults (in the worst case, not to represent a loss of communication to all SSs in the cluster). The obvious solution, at the expense of more complexity and increased investment, is to duplicate the backbone connection at a different SS in the cluster and also to establish hot-standby mechanisms that allow alternative backbone connections to temporarily assume the role of the master when there is a problem at the primary backbone connection.

As an example, backbone and BPL master faults can be solved with redundant backbones and redundancy L3 protocols such as Virtual Router Redundancy Protocol (VRRP) (in RFC 5798).

7.3.2 Narrowband PLC Telecommunications over Low-Voltage Grid

NB PLC deployments are commonly used for smart grid applications over LV networks. In this grid segment, the advantages provided by these technologies are maximized due to the diversified, extended, and often unknown details of

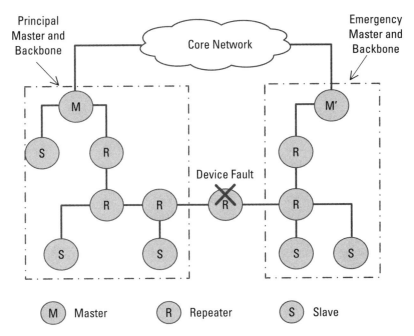

Principal Master and Backbone

Emergency Master and Backbone

Core Network

Device Fault

M Master R Repeater S Slave

Figure 7.10 BPL link fault scenario.

the LV grid. NB PLC is specifically used for massive smart metering and LV grid control and automation.

Chapter 5 covered the wide catalog of technologies classified as NB PLC; this section is based on PRIME. Other technologies could have been also used for the study with possibly similar results, but PRIME has been chosen because it has been widely and successfully deployed in several utilities, being the new generation standards-based NB PLC technology with the highest number (tens of millions) of installed devices [27].

7.3.2.1 Advanced Metering Infrastructure (AMI) Based on NB PLC ITU-T G.9904 PRIME

Architecture

The most traditional AMI (smart metering) architecture is shown in Figure 7.11. The MDMS is shown as an entity that connects to a head end (HE) that manages the smart meter access with the data concentrators (DCs) in the SSs. From the DCs to the HE and the HE to the meter data management system (MDMS), an IP broadband telecommunication solution is used (see Chapter 6 for details on all the possibilities and Chapter 9 for a specific use case). Over these two interfaces, standard application level protocols are used to exchange metering information [namely Web services and File Transfer Protocol (FTP)]. For example, Web services are published by the DC and used by the HE; FTP is

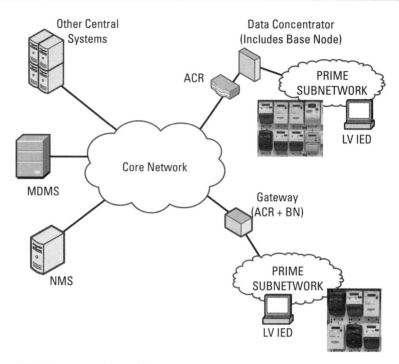

Figure 7.11   Smart metering architecture.

used to send XML (Extensible Markup Language) files with the metering data from the DC to the MDM. At the DC level, the Application Layer Protocol and object model used is DLMS/COSEM [29].

A variant of the DC-based architecture can be offered if the Application Layer Protocol (DLMS/COSEM) is integrated in the HE. Thus, the DC would be replaced by a telecommunication gateway (GW) managed from the HE. This is shown in Figure 7.11.

PRIME can be used both in the traditional DC or the new GW-based architecture. PRIME works in both cases on the LV grid to connect smart meters through PLC. PRIME subnetwork components are the BN and the SNs; the BN which is the master of the subnetwork, providing connectivity to every smart meter electrically connected to it and equipped with a PRIME SN (SN can behave as terminals or switches).

BNs are integrated in DCs or GWs, and are typically installed at the SSs; this is the typical location for BNs, although nothing prevents their installation closer to the smart meters (e.g., inside high-rise buildings to increase the bandwidth available for the smart meters).

At PRIME level, the BN establishes connections with the SNs to transport the application layer traffic between the smart meters and the DC or GW. In the case of DLMS application with non-IP smart meters (the vast majority of the smart meters deployed for massive smart metering), IEC-61334-4-32 connections are used, although PRIME supports convergence layers for IPv4 and IPv6.

Parametrization

PRIME can adapt to different applications and grid circumstances. The standard defines several parameters for this; some are dynamically adjusted based on channel response, while others are implementation decisions. For the use case, the most appropriate configuration was found to be as defined by the smart metering profile in [30].

- BNs:
  - CFP not used, if smart metering is the only application running in the use case, and all the smart meters have the same priority.
  - Beacon slot allocation avoiding switches in the same link to use consecutive beacon slots to avoid problems with low-performance SNs.
  - Mandatory ARQ (automatic repeat request), as in annex I of [30]: Maximum window size ("winsize" parameter) of 6, a maximum of 5 retries, 1 second of delay between retries, and piggybacking of the acknowledgment (ACK) in data frames.
  - Single-phase injection: three-phase systems perform better if injection is made between single-phase and neutral wires.
- SNs: ARQ optional; if used, same parameters as the BN.
- BNs and SNs:
  - Most robust scheme (DBPSK with FEC).
  - PRM disabled, to improve performance in interoperability conditions.
  - Maximum SAR size of 64 bytes, as a compromise between data performance and periodic noise impact [31].

Performance Management

Performance monitoring tools are needed both during the deployment phase and in operation. The set of tools used during deployment should be oriented to help in the commissioning process, ensuring correct network performance for any of the supported applications. Once the networks are operational, some additional tools are needed to detect potential problems promptly.

In the network deployment phase, a performance assessment needs to be done on each deployed subnetwork, as each SS LV grid is different. This performance assessment is described in [31] to study network availability and stability. The BN is configured to send any topology change for one week. These changes are registered in a log file and sent to the NMS. At the same time, application traffic is simulated to check if there is an influence on the network stability. This traffic is simulated in the BN to retrieve metering information from every smart meter in the subnetwork in a cyclic way. The information retrieved should be equivalent to the daily reports used by the MDMS (if there are several report types, groups of 2 days should be allocated for each report). Each data request cycle over a smart meter is registered (successfully received data or not), and sent to the NMS for further analysis. Figure 7.12 shows a sample of such a performance analysis. As a result of this process, there will be some recommendations to be made in order to improve the performance of the subnetwork: replacement of devices on failure, special parametrization of the BN, mitigation of noise disturbances in the grid, bug detection, and so forth.

In the operational phase, as the subnetworks are being used by the LV smart grid applications (e.g., smart metering, LV grid monitoring), the applications will have their own performance monitoring tools and indicators. Eventual problems will happen and further investigation at telecommunication level (PLC) will be needed. Apart from this reactive approach, PRIME is a PLC telecommunication technology that can be managed at its own telecommunication level with a NMS.

A SNMP-based NMS can retrieve some of the MIB objects in the BNs periodically [32] (topology of the network, convergence layer connections, channel occupation, and so forth) to perform statistical analysis on them in order to generate alarms on particular subnetworks if certain unusual conditions are detected. An example is shown in Figure 7.13, a graph comparing the typical traffic pattern of a network with the traffic of a particular day, can give a view of the distribution of the traffic or even detect congestions generated by an unexpected application procedure.

These analyses need also to be complemented with sniffing capabilities in the BNs. The BN will encapsulate each PRIME PDU transmitted or received into a TCP/IP packet and send it to the NMS so that a graphical detailed analysis can be performed.

7.3.2.2   Narrowband PLC for Low Voltage Grid Control

Introduction

PLC-based smart metering deployments can provide additional smart grid functionalities if properly chosen and deployed. These applications extend the traditional remote control capabilities already present in the HV and MV segments, to the LV grid.

The kinds of applications that can take advantage of the smart metering infrastructure are categorized in two groups:

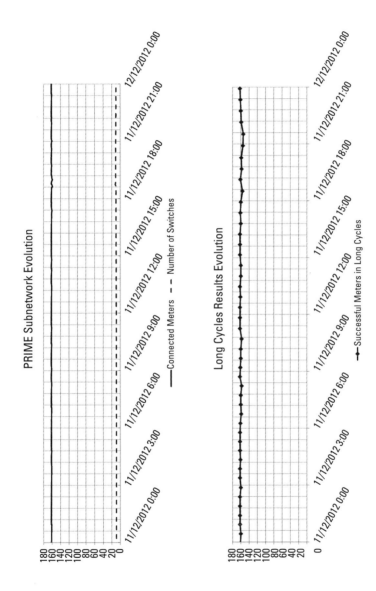

Figure 7.12 Performance assessment results during deployment.

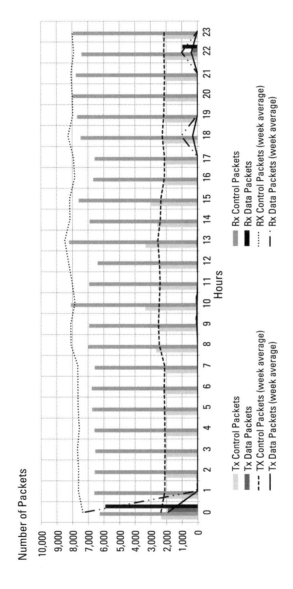

Figure 7.13   PRIME subnetwork evolution managed from the NMS.

- Applications leverage metering information to analyze the LV network and improve the knowledge and operations of the grid;
- Applications leverage telecommunication infrastructure already deployed to transport other traffic different from metering.

In the first group, applications such as LV supervision, fraud detection, and line and phase identification are available. In the second group, applications such as remote automation of the LV grid are included (i.e., LV DA).

Metering Data Utilization

LV grid utility databases are not as accurate as the rest of the voltage levels [33]. Considering that the real connectivity of the smart meters' points of supply is very valuable for any LV operation, smart metering data is useful to capture connectivity directly from the field.

For this purpose, the first step in the connectivity identification is the transformer to which a meter is connected. When using PRIME PLC technology, as the channel is the LV feeder itself, if a smart meter is registered to a BN installed in an SS, in a context where all SSs are deployed with BNs, the smart meter electrically connected to a SS has a high probability (certainty can be achieved measuring power level) of being connected to this SS. The next step is the identification of the feeder and phase in the SS. Feeder identification is done by sensing the signal power received from a certain smart meter through each feeder at the SS. Phase identification is a built-in mechanism of PRIME.

Another application derived from the utilization of the metering data is the LV supervision, in order to detect power outages or inefficiencies in the grid (e.g., phase unbalance in the LV side of the transformer). Prior to the deployment of the smart metering infrastructure, the only way to detect an outage in the LV network was customer complaints. Inefficiencies in the LV grid were detected with expensive field measurement campaigns. If, along with a smart metering deployment, an LV header meter is installed at the SS to register both current and voltage, DCs or central systems can detect outages or unbalances in real time.

Finally, fraud detection mechanism can be implemented comparing the information gathered by the LV header meter with the sum of all powers registered by the smart meters connected to this transformer. If the difference is higher than the expected technical losses, there is potential fraud that can be traced and finally checked on field.

Smart Metering Telecommunication Infrastructure Utilization

The PLC infrastructure deployed for smart metering can be used for other applications, due to its multiservice nature and its capability to transport IP traffic. LV IEDs can be deployed and connected with central control systems through this end-to-end connectivity through the local, access, and core networks. For example, LV electrical switches are used to dynamically connect a particular LV feeder of one MV/LV transformers in an SS to another one in another SS to give an electricity service backup for customers.

PRIME PLC specification defines an IP convergence layer to be used over the PLC MAC and PHY and also defines a CFP to reserve bandwidth for critical applications. Using those two functions, a LV IED connected to the electrical switch can be reached from the central control application (SCADA), using any remote control protocol based on IP such as IEC 60870-5-104.

References

[1]   Schwartz, M., "Carrier-Wave Telephony over Power Lines: Early History," *IEEE Communications Magazine*, Vol. 47, No. 1, 2009, pp. 14–18.

[2]   Dressler, G., *Hochfrequenz-Nachrichtentechnik für Elektrizitätswerke*, Berlin: Julius Springer, 1941.

[3]   Ferreira, H. C., et al., (eds.) *Power Line Communications: Theory and Applications for Narrowband and Broadband Communications over Power Lines*, New York: Wiley, 2010.

[4]   Razazian, K., et al., "G3-PLC Field Trials in U.S. Distribution Grid: Initial Results and Requirements," *Proc. 2011 IEEE International Symposium on Power Line Communications and Its Applications*, Udine, Italy, April 3–6, 2011, pp. 153–158.

[5]   Mak, S., and D. Reed, "TWACS, a New Viable Two-Way Automatic Communication System for Distribution Networks. Part I: Outbound Communication," *IEEE Transactions on Power Apparatus and Systems*, Vol. 101, No. 8, 1982, pp. 2941–2949.

[6]   Mak, S., and T. Moore, "TWACS, A New Viable Two-Way Automatic Communication System for Distribution Networks. Part II: Inbound Communication," *IEEE Transactions on Power Apparatus and Systems*, Vol. 103, No. 8, 1984, pp. 2141–2147.

[7]   Nordell, D. E., "Communication Systems for Distribution Automation," *Proc. Transmission and Distribution Conference and Exposition 2008*, Chicago, IL, April 21–24, 2008, pp. 1–14.

[8]   HomePlug Alliance, "About the HomePlug® Alliance," 2016. Accessed March 24, 2016. http://www.homeplug.org/alliance/alliance-overview/.

[9]   HD-PLC Alliance, "Welcome to HD-PLC Alliance." Accessed March 24, 2016. http://www.hd-plc.org/modules/alliance/message.html.

[10] Oksman, V., and J. Egan, eds., *Applications of ITU-T G.9960, ITU-T G.9961 Transceivers for Smart Grid Applications: Advanced Metering Infrastructure, Energy Management in the Home and Electric Vehicles*, Telecommunication Standardization Sector of International Telecommunication Union (ITU-T), 2010. Accessed March 24, 2016. https://www.itu.int/dms_pub/itu-t/opb/tut/T-TUT-HOME-2010-PDF-E.pdf.

[11] Lampe, L., A. M. Tonello, and T. G. Swart, *Power Line Communications: Principles, Standards and Applications from Multimedia to Smart Grid*, 2nd ed., New York: Wiley, 2016.

[12] Goodwins, R., and R. Barry, "Plug into the Net," *ZDNet*, 1998. Accessed March 26, 2016. http://www.zdnet.com/article/plug-into-the-net/.

[13] Sendin, A., "PLC Commercial Deployment of Iberdrola," *Proc. 2004 IEEE International Symposium on Power Line Communications and Its Applications*, Zaragoza, Spain, March 30–April 2, 2004.

[14] Abad, J., et al., "Extending the Power Line LAN Up to the Neighborhood Transformer," *IEEE Communications Magazine*, Vol. 41, No. 4, 2003, pp. 64–70.

[15] Galli, S., O. Logvinov, "Recent Developments in the Standardization of Power Line Communications Within the IEEE," *IEEE Communications Magazine*, Vol. 46, No. 7, 2008, pp. 64–71.

[16] Oksman, V., and S. Galli, "G.hn: The New ITU-T Home Networking Standard," *IEEE Communications Magazine*, Vol. 47, No. 10, 2009, pp. 138–145.

[17] Latchman, H. A., et al., *Homeplug AV and IEEE 1901: A Handbook for PLC Designers and Users*, New York: Wiley-IEEE Press, 2013.

[18] Zaballos, A., et al., "Survey and Performance Comparison of AMR over PLC Standards," *IEEE Transactions on Power Delivery*, Vol. 24, No. 2, 2009, pp. 604–613.

[19] Liu, W., M. Sigle, and K. Dostert, "Channel Characterization and System Verification for Narrowband Power Line Communication in Smart Grid Applications," *IEEE Communications Magazine*, Vol. 49, No. 12, 2011, pp. 28–35.

[20] Biglieri, E., "Coding and Modulation for a Horrible Channel," *IEEE Communications Magazine*, Vol. 41, No. 5, 2003, pp. 92–98.

[21] Hrasnica, H., A. Haidine, and R. Lehnert, *Broadband Powerline Communications Networks: Network Design*, New York: Wiley, 2004.

[22] Berganza, I., et al., "PRIME On-Field Deployment: First Summary of Results and Discussion," *Proc. 2011 IEEE International Conference on Smart Grid Communications*, Brussels, October 17–20, 2011, pp. 297–302.

[23] PRIME Alliance, "Alliance," 2013. Accessed March 24, 2016. http://www.prime-alliance.org/?page_id=2.

[24] G3-PLC Alliance, "About Us." Accessed March 24, 2016. http://www.g3-plc.com/content/about-us.

[25] Sendin, A., R. Guerrero, and P. Angueira, "Signal Injection Strategies for Smart Metering Network Deployment in Multitransformer Secondary Substations," *IEEE Transactions on Power Delivery*, Vol. 26, No. 4, 2011, pp. 2855–2861.

[26] Sendin, A., et al., "Strategies for PLC Signal Injection in Electricity Distribution Grid Transformers," *Proc. 2011 IEEE International Symposium on Power Line Communications and Its Applications*, Udine, Italy, April 3–6, 2011, pp. 346–351.

[27] bmp Telecommunications Consultants, *Worldwide Broadband PLC Atlas 2016*.

[28]   Solaz, M., et al., "High Availability Solution for Medium Voltage BPL Communication Networks," *Proc. 2014 18th IEEE International Symposium on Power Line Communications and Its Applications*, Glasgow, March 30–April 2, 2014, pp. 162–167.

[29]   DLMS User Association, "What Is DLMS/COSEM." Accessed March 23, 2016. http://www.dlms.com/information/whatisdlmscosem/index.html.

[30]   PRIME Alliance Technical Working Group, *PRIME v.1.3.6 Specification*, PRIME Alliance, 2013. Accessed March 26, 2016. http://www.prime-alliance.org/wp-content/uploads/2013/04/PRIME-Spec_v1.3.6.pdf.

[31]   Sendin, A., et al., "Strategies for Power Line Communications Smart Metering Network Deployment," *Energies*, Vol. 7, No. 4, 2014, pp. 2377–2420.

[32]   Sendin, A., et al., "Adaptation of Powerline Communications-Based Smart Metering Deployments to the Requirements of Smart Grids," *Energies*, Vol. 8, No. 12, 2015, pp. 13481–13507.

[33]   Marron, L., et al., "Low Voltage Feeder Identification for Smart Grids with Standard Narrowband PLC Smart Meters," *Proc. 2013 7th IEEE International Symposium on Power Line Communications and Its Applications*, Johannesburg, March 24–27, 2013, pp. 120–125.

8

Radio Technologies Supporting the Smart Grid: Use Cases

Radio as a telecommunication medium is of great interest in large-scale and widespread deployments, as radio-based networks provide for faster network implementations and broader coverage than wireline alternatives. This is usually associated with lower infrastructure costs for the same coverage.

However, radio is a container concept for different technologies. We refer to radio in general, even if the frequency spectrum has different properties in terms of propagation, penetration, cost per unit of the bandwidth, and availability. Radio systems exhibit multiple differences across world regions and countries in terms of regulations, allotted frequencies, and standards. When evaluating the suitability of the different radio systems, the specifics of each scenario and available solutions have to be analyzed considering the needs of the smart grid, as radio systems are very different among them.

There are a number of elements that become relevant in the analysis of radio for smart grids:

- Public mobile radio services versus private radio networks: This consideration applies very specifically to radio, as many of the most popular radio systems today are targeted to residential customers with no smart grid-specific orientation.

- Standard versus proprietary solutions: The availability of license-exempt bands has stimulated the development of multiple proprietary solutions; this by itself clearly shows there are many different ways that radio can effectively support quick access to telecommunication services. While proprietary solutions avoid the complexity of compatibility certification between competing vendors, they also exhibit limitations that advise the

229

users of critical industrial and utility applications to consider them very cautiously or even avoid them [1].

- Full- or partial-coverage radio systems: Point-to-multipoint, point-to-area, or mesh radio systems usually cover limited areas around their base stations, connected through radio and nonradio technologies. Cellular systems tend to provide territory wide coverage; on the contrary, short-range device (SRD) radio networks are typically used in the edges of any network to connect close-standing elements.

Real telecommunication systems are heterogeneous and mix wireless and wireline technologies depending on the specific needs, purposes, and opportunities in the various parts of the network. It is unusual to find a telecommunication system that is based only on radio technologies.

8.1 Architectures

Considering the variety of radio technologies and systems available, there is no single architecture for radio-based smart grid deployments. Different topologies can be identified:

- Point-to-point;
- Mesh;
- Point-to-multipoint/point-to-area;
- Cellular land mobile;
- Satellite.

Topologies cover the different ways in which radio technology can be deployed and fit better specific purposes, frequency ranges, or scenarios. All of them can be mixed to create complex structures.

8.1.1 Point-to-Point Radio

Radio transmission completes wireline transmission media where point-to-point connectivity cannot be easily implemented with physical wires. Radio links (as point-to-point radio instances are called) can cover distances up to 100 km and more, depending on the frequency, the transmit power, the receiver sensitivity, and the antennas used. The cost of a radio link is typically lower than its equivalent physical wireline alternative, independent of the distance to be covered.

Point-to-point radio can be used at core and access network levels for transport purposes. Table 8.1 shows a high-level representation of frequencies, distances, and bandwidths for typical commercially available systems.

A radio link (Figure 8.1) can be used to connect a primary substation (PS) to the telecommunication network. A chain of radio links (radio links sequentially connected) has to be deployed if distance or nonline-of-sight (NLOS) conditions do not allow for a direct connection. A radio link can be also used to connect radio base stations that configure point-to-multipoint/area wireless coverages. These radio base stations are located at hills, mountain tops, or building rooftops, and radio links are a cost-effective, quick-to-deploy option.

8.1.2 Mesh Radio

Mesh radio (Figure 8.2) is a network topology that configures networks based on a high number of point-to-point connections among nodes, where each node relays data for other nodes in the network. In wireless mesh, radio is used as the transmission medium, while other L2 (switching) and L3 (routing) protocols manage the way that each node connects the various network paths. Mesh radio tends to be associated with self-healing networks, although this is not inherent to the mesh topology, but to the capabilities of the switching/routing protocols if there are redundant routes among nodes.

Table 8.1
Characteristics of Radio Links in the Different Frequencies

Frequency Band (GHz)	Range	Capacity	Use (Range – Capacity – Investment)
2	Long (100 km)	Medium (2, 2 × 2 Mbps)	L-M-Lw
4	Long (100 km)	Medium (2, 2 × 2 Mbps)	L-M-Lw
6	Long (80 km)	High (34, 2 × 34 Mbps)	L-Hi-Hi
7/8	Long (80 km)	High (34, 2 × 34 Mbps)	L-Hi-Hi
11	Medium (40 km)	Medium (4 × 2, 8 × 2 Mbps)	M-M-M
13	Medium (40 km)	Medium (4 × 2, 8 × 2 Mbps)	M-M-M
15	Medium (30 km)	Medium (4 × 2, 8 × 2 Mbps)	M-M-M
18	Medium (20 km)	Medium (4 × 2, 8 × 2 Mbps)	M-M-M
23	Medium (15 km)	Medium (8 × 2, 16 × 2 Mbps)	S-Hi-M
26	Short (10 km)	Medium (8 × 2, 16 × 2 Mbps)	S-Hi-M
38	Short (5 km)	High (34 Mbps, STM-1)	S-Hi-M

Source: [2]. (L = Long; M = Medium; S = Short; Hi = High; Lw = Low.)

Figure 8.1   Radio links in electricity premises.

Radio Mesh Network

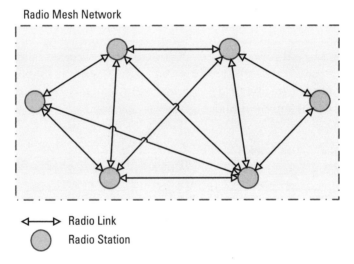

◄———► Radio Link

    Radio Station

Figure 8.2   Mesh radio network.

Mesh radio networks are often used in short-range coverages and/or sub-gigahertz frequency scenarios.

8.1.3   Point-to-Multipoint/Point-to-Area Radio

Radio, through the proper antennas, can create a coverage area where multiple end-points can be served from one single central point. Coverage area is defined by ITU-R Recommendation V.573-5 as the "area associated with a transmitting station for a given service and a specified frequency within which, under specified technical conditions, radiocommunications may be established with one or several receiving stations."

The central location from which coverage is created is called the radio base station or base station [other names are, e.g, radio repeater station, base transceiver station (BTS), node-B]. Radio base stations are defined as the installation intended to provide access to a telecommunication system by means of radio waves.

This type of radio telecommunication systems uses specific high-altitude locations to create large coverage footprints, for any location in the service area (e.g., with ranges of 30 km of the radius around the base station, an area of 500,000 km^2 could be theoretically covered with around 200 base stations).

Calculations based on line of sight (LOS) can be used for point-to-multipoint coverages, as they are essentially point-to-point links from the base station to a discrete number of end-points in fixed locations. If coverage needs to be extended to mobile elements, calculations need to be based on mobile propagation prediction methods.

Figure 8.3 shows typical representations of point-to-multipoint and point-to-area radio networks.

8.1.4 Cellular Land Mobile Radio

Cellular radio refers to modern [i.e., newer than the 1980s; from first generation (1G) systems onwards] networks mainly used for public mobile telecom-

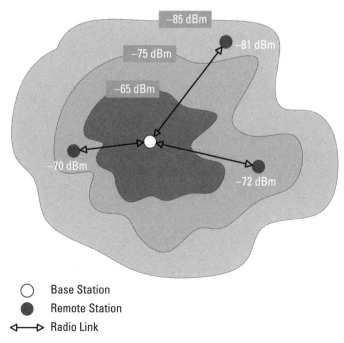

Figure 8.3   Point-to-multipoint and point-to-area coverages.

munications. Cellular radio, in terms of standards, is different in the three International Telecommunication Union (ITU) world regions and although the basic underlying technologies are similar, the systems are incompatible.

Cellular systems are similar to point-to-area radio systems. The cellular concept dates back to 1947 [3, 4], and it was not massively applied until public mobile systems became popular. The original problem laid in the saturation of the radio network deployed with a point-to-area coverage concept, and the limitation of the frequency spectrum channels. Cellular systems lead to the creation of an increasing number of smaller cells, each having a base station, able to always reuse the same set of frequencies in a growing smaller area. The bandwidth assigned to each base station is continuously reused in cells, adapted to the density of mobile radio users in the area.

Examples of cellular systems are those of the successive 1G, second generation (2G), third generation (3G), and fourth generation (4G) technologies, with different standards in the different regions of the world. These systems have progressively allowed for the use of higher frequencies, and now the trend is to reuse the lower ones that are becoming available due to the obsolescence of the oldest systems. Figure 8.4 shows typical cellular antennas.

8.1.5   Satellite Radio

Satellite is the paradigm of global coverage. Satellites can be used as "bent pipes" in the sky, connecting locations in the Earth that otherwise would experience LOS problems. This principle is used to connect the ground with satellite base stations that provide broad coverage footprints on the Earth, offering connectivity to remote areas.

Satellites can be placed in different orbit types. There are low Earth orbit (LEO), medium Earth orbit (MEO), and geostationary Earth orbit (GEO) satellites, with different characteristics in terms of provided services, the telecommunication parameters needed (e.g., power or antenna gain), and constraints

Figure 8.4   Cellular antenna colocated in a primary substation.

imposed on the service locations (e.g., the size and orientation of the antennas). Moreover, the position of the satellites determines their lifespan and, consequently, the need to reinvest in new elements of the telecommunication platform.

Satellites serve today probably as many applications as any other land system, including mobile communications similar to cellular networks. Satellites are like mobile base stations on Earth, but providing broad regional coverage from their privileged high-altitude location.

Satellite systems are as conceptually attractive as expensive to implement. The cost of launching and placing satellites in orbit and the time that elapses from the initial design until the satellites are operational usually involve high risks. Furthermore, the distances involved (requiring very accurate transmitters and receivers), the non-negligible delays in communications, and the low throughput (as they need to be shared among a large number of customers) make satellite systems a not-always attractive solution. Operational limitations include LOS conditions with the satellite (a challenge in urban areas with high-rise buildings), and terminals requiring large antennas. Figure 8.5 shows a satellite terminal installation inside a PS.

8.2   Design Guidelines

This section covers the aspects to be considered in the design of radio-based networks, relevant to understand private network design and public radio services constraints.

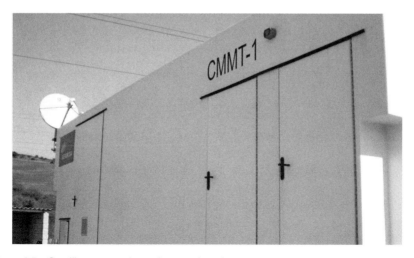

Figure 8.5   Satellite antenna in a primary substation.

8.2.1 Frequency Bands

The key element to a radio system design is the availability of frequencies, organized in channels (ITU-R V.662-3 defines the term frequency channel as the "part of the frequency spectrum intended to be used for the transmission of signals and which may be defined by two specified limits, or by its centre frequency and the associated bandwidth, or by any equivalent indication"; the radiocommunication vocabulary definitions are now covered by ITU-R V.573-6, that refers to [5]). The frequency defines the distance to be covered (for a certain transmit power and receiver sensitivity) and the available bandwidth.

The spectrum usage depends on the multiple access scheme to be used: frequency division multiple access/time division multiple access (FDMA/TDMA) or code division multiple access (CDMA). A classical frequency channel is made of a transmission and a reception pair of frequencies (i.e., a full duplex FDMA channel). However, channels can also have one single frequency (used in a shared manner to transmit and receive at different times), or even share the same frequency and time resources with CDMA thanks to different channel codes. In the discussion that follows, we will refer to classical two-frequency channels if not stated otherwise.

Frequency bands are a scarce resource, managed by the different national administrations in their territories, and coordinated globally by the ITU allocations. Administrations [e.g., the Federal Communications Commission (FCC) in the United States, the Office of Communications (OFCOM) in the United Kingdom, and so forth] provide licenses for use at national level, within the recommendations of the ITU. The spectrum assigned by these administrations is licensed and can only be used by the licensee under the conditions stated in the license. Any illegitimate use of this assigned spectrum by a nonlicensee is protected by that administration.

There are examples of radio allocations for private point-to-point radio and for systems with global implementations such as Global System for Mobile (GSM). The following considerations cover how these frequency allocations have global and local legal protection:

- Point-to-point links working in specific bands are the example of channel band plan definition that serves the purpose of allowing vendors to design products that, following the ITU-R recommendations, can be used globally. For example, the 7 GHz microwave band is widely used for point-to-point telecommunications in tens of kilometers with tens of megabits per second of capacity; there are different possible channels arrangements, and each national administration can select the most appropriate (see Figure 8.6).

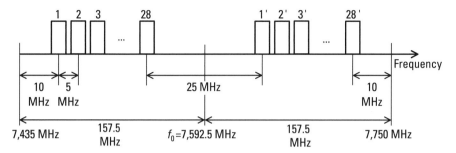

Figure 8.6 Valid ITU-R F.385-10 channel arrangement.

- GSM systems are an example of a global allocation: 880 to 915 MHz and 925 to 960 MHz bands (allotted in some countries) are reserved in accordance with the EC decisions 2009/766/CE modified by the 2011/251/UE. ITU refers to IMT ITU-R M.1036 in RR 5.313A and 5.317A.

The way that the two frequency uses are assigned to end-users depends on the specific country. For point-to-point assignment, a process to obtain licenses on an individual link basis is usually in place; for licenses related to public systems for commercial use, there is typically either a competitive tender or an auction process.

8.2.1.1 License-Exempt Frequency Bands

There are a few frequency bands that are reserved by the ITU worldwide for specific nontelecommunication purposes and that can be used without any license. The definition of this specific application is known as ISM (Industrial, Scientific, and Medical) and it is made in the ITU-R RR as adopted by the World Radiocommunication Conferences (WRCs). ISM applications (of radio frequency energy) cover "operation of equipment or appliances designed to generate and use locally radio frequency energy for industrial, scientific, medical, domestic or similar purposes, excluding applications in the field of telecommunications" (see 1.15 in [7]).

The details of the bands under the ISM denomination are given in 5.150, 5.138, and 5.280 [7]. The most popular ISM frequencies are those with center frequencies of 915 MHz, 2.45 GHz, and 5.8 GHz, although others exist.

Frequency bands for ISM applications must accept any harmful interference coming from devices and other users working in the same band and are affected by other limitations (mainly transmit power) that administrations define for these services not to affect other licensed users in nearby bands. For example, for the 2.4 GHz band, OFCOM in [7] sets a maximum EIRP of 100 mW for wideband transmission systems between 2.4 and 2.4835 GHz.

ISM bands are not the only ones that can be used in a license-exempt basis. The Report ITU-R SM.2153-5 covers the use of the SRDs for "radio transmitters which provide either unidirectional or bidirectional communication and which have low capability of causing interference to other radio equipment." SRDs are not considered a radio service under the ITU RR, and thus cannot have allocations. However, the different regions and countries recognize that SRD systems are useful and should be given a proper consideration. As an example (more in ITU-R SM.2153-4), both the European Commission (EC) and the European Conference of Postal and Telecommunications (CEPT) produced a mandatory harmonization decision (2006/771/EC), specific technology based decisions (ECC/DEC(06)04), and a recommendation (CEPT ERC/ Rec 70-03) intended to "promote the most efficient implementation of these devices" [8].

SRD devices are "permitted to operate on a non-interference and non-protected basis," although licenses could apply to some of them at a national level. Telecommand, telemetry, alarms, and radio frequency identification (RFID) are identified as subject of SRD. The ITU-R Recommendation SM.1896 recommends the frequency ranges that should be used for a global or regional harmonization (7,400–8,800 kHz; 312–315 MHz; 433.050–434.790 MHz; 862–875 MHz; 875–960 MHz).

The use of unlicensed or license-exempt bands is not recommended for critical smart grid services, as the investment could be compromised by unexpected interference for which there is no specific protection.

8.2.2 Radio Propagation

Propagation is a key consideration to predict how far and how reliably a radio signal can travel within the given constraints of a system. Propagation in point-to-point systems is studied from the perspective of the terrain (the obstacles in the path between transmitter and receiver that attenuate the radio signal) and the atmospheric effects (troposphere and hydrometeors [10] such as rain and ice). Propagation studies in point-to-area systems (either with mobile or stationary receivers) include many other probabilistic aspects, given the unpredictability in channel response that mobility implies. Cellular systems take the propagation calculations to their full extent, because they are heavily applied to urban areas with many reflections, obstructions, and diversity of effects with shorter distances and higher frequencies. Radio propagation is in difficulty if waves need to traverse floors, walls, and cellars.

As a result of propagation studies, designers can predict within certain probabilistic values what the received signal levels will be in the different conditions. Propagation studies take into consideration frequency, bandwidth, transmit power, sensitivity of receivers, services, and number of users. Figure 8.7

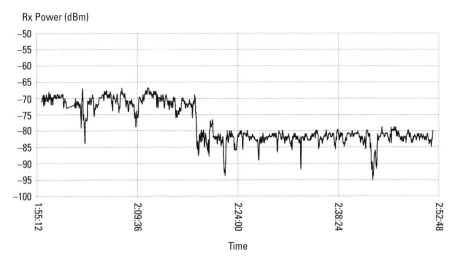

Figure 8.7 Received signal strength in a fixed receiver in a 2.5G GPRS system.

shows the received signal strength of a 2.5G GPRS network, where received signal fluctuations can be observed in static receivers.

ITU-R and specifically Study Groups 1 (Spectrum Management; ITU-R SM Series) and 3 (Radiowave Propagation; ITU-R P Series) produce important information as recommendations for radio propagation analysis and prediction.

8.2.3 Bandwidth

Radio Regulations (RRs) allocate certain frequency ranges for different applications within specific start and stop frequencies. The way this frequency range or band is shared among different users is through the channel concept. Every spectrum user in a certain band can use one or more channels. The bandwidth is enclosed within the channel space at a frequency band (ITU-R V.662-3 [5] defined bandwidth as "the quantitative difference between the limiting frequencies of a frequency band"). Within this channel, the information has to be conveyed using different multiplexing and/or modulations. Thus, in order to configure a point-to-area or point-to-point radio network, channels must be assigned to the different base stations. Similarly, if a set of point-to-point radio links (including mesh topologies) needs to be established, radio channels must be coordinated in the different radio links (see Figure 8.6).

Channel bandwidth is fundamental to determine the throughput of a certain radio network. For example, in a point-to-area system using the band around 450 MHz, typical channel bandwidths are usually around 12.5 or 25 kHz. With simple modulation techniques, data rates of nearly 20 kbps can be achieved; depending on the link budget, higher-order modulations could be used for higher spectral efficiency. This would be the maximum capacity to be shared by all the users/locations served by a single base station with one channel, when the system is based on point-to-area coverage. The example of a typical cellular GSM system with 2.5G capacity within the 900 MHz frequency band and 200 kHz channels provides a throughput of 270 kbps to be shared by all the users under the coverage of the base station. If throughput needs are higher, and no other frequency can be found, either the system uses its cellular characteristics and introduces more channels through reuse, or it finds more channels to be used (usually difficult). The 3G-based systems use channel bandwidths of 5 MHz [e.g., universal mobile telecommunications system (UMTS), using wideband code division multiple access (WCDMA)], which allow for higher data rates.

Spectrum availability increases with frequency. However, both the limitations of electronics and, above all, the limitations of propagation as frequency increases make it difficult to use the higher part of the spectrum for certain applications (this is one of the reasons why 3G technologies intend to use spectrum previously allocated to 1G and 2G technologies).

Finally, a mention must be made to the trade-off between bandwidth and propagation; the larger the bandwidth, the shorter is the propagation range, as the noise captured by the receiver is higher.

8.2.4 Radio Base Stations

Radio base stations are the nodal points of coverage-based radio networks. Radio base stations are the sites used to distribute and collect radio transmissions in the area where the service is provided.

Base stations are of different types depending on the architecture and the frequency in the system. The architecture determines how the base station interacts with the rest of the elements in the system (base stations and endpoints). Frequency constrains the size of the radiating elements (i.e., antennas) in the system, and the distances between different components (i.e., the higher the frequency, the smaller the radiating elements and the shorter the distances).

Radio base stations are comprised of radio-related elements and other components:

- Shelter: Electronic equipment need to be protected from weather conditions, so either these devices are by design prepared to work in specific outdoor conditions (e.g., compliant to IEC 60529 or other references such as the American NEMA Standard Publication 250 [11]) or need sheltering to provide controlled environmental and access conditions.

- Tower: The tower is the mechanical structure where the antennas will be installed. It also supports cables and waveguides connecting the antennas to the radios. The tower height depends on coverage needs, surrounding terrain and distance to end-points.

- Power supply: Radios can be fed in ac or, more typically, dc. Power supply needs to be taken from the grid. Radios need to work also in the event of a power failure, so usually batteries exist. These batteries are dimensioned to last for a certain number of hours (e.g., mobile cellular can be dimensioned for a few hours' duration [12] while mission-critical services may require 24 to 48 hours back up time [13, 14]). Additional local ac generation may exist.

- Access: The installation of electronic devices, tower, antennas, and shelters requires the transport of materials during the construction period. Its periodic maintenance needs also to be secured even in hard weather conditions.

There are different locations for radio base stations:

- On mountains or hills;
- On the top of high-rise buildings;
- On the walls of buildings;
- On low-rise constructions;
- On transmission towers or pylons;
- On poles;
- On satellites;
- On mobile vehicles.

Space constraints, accessibility, power availability, and visibility are different in each case and need to be considered in base station design, construction, and maintenance.

8.2.5 Network Design

Radio network design and frequency planning consists of creating a detailed plan with the optimal network solution, which includes selecting the radio base station sites and assigning different frequency channels to each of them. The services to be delivered, their quality, the area to be served and any other constraint (e.g., required standards to follow and frequency band to use) are inputs for this process.

Radio network design has two main parts: dealing with the range/coverage design, and the traffic calculations. Range/coverage design deals with the calculation of the received signal levels and the transmit power control to avoid interferences; traffic calculations are intended to guarantee that the system will cope with the traffic generated by its users.

Radio networks consisting of multiple individual point-to-point links (including point-to-multipoint and meshed networks) are designed with different propagation models than networks with point-to-area configurations (including cellular). ITU-R Recommendation P.1144 offers useful guidelines for the use of the different propagation methods.

Networks that are not point-to-area are in general easier to design since locations are fixed and signal levels are more predictable and deterministic. Propagation studies are undertaken with accurate methods based on physics (Fresnel ellipse, knife-edge diffraction models, and so forth). The location of the base stations can be generally selected to avoid terrain obstacles, so that LOS can be achieved. Traffic calculations are usually simple as the traffic matrixes (inputs and outputs at different points in the system) are known in advance. In mesh systems, what-if scenarios must be considered to guarantee that traffic can be admitted in all network conditions.

The case of the point-to-area needs to be studied with propagation models based on measurements, such as the ITU-R Recommendation P.1546 (other important models worthwhile mentioning are the Okumura-Hata [15] and the COST231 [16]). All these models predict the coverage on different hypothesis of a certain percentage of time and locations, as 100% coverage certainty is impossible. The rest of this section will cover the point-to-area case.

8.2.5.1 Coverage Design

Radio coverage range may be limited by either noise, interference, or traffic limited systems:

- Noise-limited systems are those where coverage depends on the signal-to-noise ratio (SNR). Examples are within the private mobile radio (PMR) systems for professional use.

- Interference limited systems get their range limits from the proximity of interfering sources, generally from the same system. Examples are cellular systems.

- Traffic limited systems get their range limits fixed by the amount of users and their traffic in an area. Examples are public systems and PMR trunking systems.

The system service objectives are always defined by the coverage area and the availability of the system. For example, a system may cover just the populated portions of a service area (a certain percentage of the population and a low percentage of the service territory), or it can cover the whole territory with a certain limited number of exceptions (e.g., mission-critical mobile networks). The coverage is calculated for a certain percentage of the time in the service area (50%, 90%, 95%, 99%, and so forth).

Once the service area is known, the selection of the base station sites begins. This is always an iterative process. Base station locations can be selected based on the definition of a theoretical grid with certain distances between the nodes (prospective base stations); base stations can also be selected in function of the desired coverage if the service area is not continuous or homogeneous; base stations are sometimes selected based on terrain knowledge, preexisting base stations or availability of supporting infrastructure (e.g., substations and poles).

Automated calculation tools and terrain models can be used to predict coverage levels in the service area. Once the sites (base stations) are decided in a first iteration, coverage prediction models must be applied. Various such models exist; they have evolved over time and can be classified into those which require low computational effort and few data, and those that require high computational effort and a detailed knowledge of the terrain and clutter (i.e., the representation of the mobile station detailed local environment) conditions. Radio parameters such as transmit power, sensitivity of the radios, interference tolerance, antennas, mast heights, cables, and waveguides are inputs for these calculations.

8.2.5.2 Traffic Calculations

The number of base stations and channels depends not only on the coverage but on the system data load (i.e., the traffic). Traffic calculations are especially important in mobile radio systems since users are not static. The distribution of users, although ideally uniform, varies along the day and the amount of traffic they demand is also variable.

Traffic dimensioning begins with an understanding on how many users, how much traffic, and how they occupy the different environments of the service area along the day, the different days of the week and the year (see Figure 8.8), as the patterns of use of the network are coincident with the behavioral patterns of their users. A bad traffic design will affect performance and can make a network collapse.

Traffic has been traditionally measured in Erlangs (see Chapter 2) since the main service in the systems was voice. However, modern networks are increasingly used for data purposes and traffic calculations adapt to this new circumstance by combining data network traffic calculations for peaks and averages.

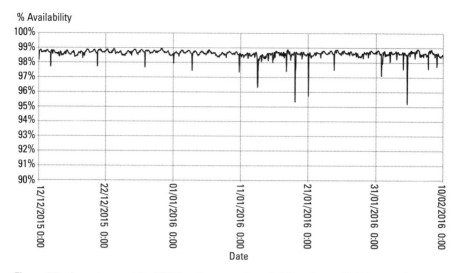

Figure 8.8   Long-term public 2G/3G radio network availability of over 20,000 secondary sub-stations.

Capacity planning uses either blocking or queuing alternatives, that is, rejecting incoming calls or holding them until there is capacity in the network to admit the traffic (Erlang B and Erlanc C are the curves used for these calculations, respectively). Blocking is the common experience in commercial networks, and queuing is the one used in many private systems to hold calls until they can be served. The target GOS completes the input data needed (defined in ITU-T Recommendation E.600 as "a measure of adequacy of a group of resources under specified conditions").

Traffic calculations determine the number of radio channels needed in each base station. However, the number of radio channels is finite and this is where the cellular concept (i.e., the channel reuse strategy) becomes useful. Early radio systems' network designs were limited by noise at the distant end (receivers). On the contrary, cellular systems with frequency reuse are limited by interference, as the base stations are close to each other. A special case of cellular systems are those based on CDMA (e.g., 2G systems such as IS-95, and 3G systems such as UMTS). In these systems, all channels share all of the band, and each channel is noise for the rest. In this circumstance, the higher the number of users, the higher the number of channels and the noise level and, consequently, the lower the SNR and the lower the range, decreasing the distance covered by the base station. This effect is known as cellular breathing; it makes base stations cover a smaller or larger area depending on the number of users and traffic in the area.

Traffic calculations are needed in order to assess the feasibility of different systems for specific traffic patterns. Reference [1] is an example of a simplified analysis for field area networks (FANs) in the context of the United States. It includes an example of calculations to determine which is the best radio communications system alternative for different smart grid service assumptions, in different environment types (urban, suburban, and rural). It examines wireless telecommunications service provider's (TSP's) network, and private network with shared spectrum, licensed or unlicensed spectrum. Although pros and cons exist for all alternatives, the public cellular networks and unlicensed radio networks are presented as the most unattractive options.

8.2.6 Quality of Service and Priority

The radio access interface is the key difference between wireline (fixed) and wireless (mobile) systems. By nature, radio is a shared medium, while most fixed lines (local loops used for telephony and xDSL) are assigned to individual users, so no user blocks the access to other users. However, in radio systems a user may be unable to gain access to the network, due to wrong resource planning or unexpected congestion.

Once access to a radio system is granted, existing resources must be managed to handle quality of service (QoS). However, once radio systems are dimensioned and deployed, their capacity is fixed and does not allow substantial modifications. This is crucial in public networks that are reused for other purposes different to those for which they were dimensioned. If there is no way to give priorities to some users above the rest, no QoS can be guaranteed.

Schemes for privileged access to public commercial networks have been recently adopted for voice services (WPS [16] in the United States and MTPAS [17] in the United Kingdom). Reference [18] described how it is technically achieved for CDMA-based systems. Long-term evolution (LTE) has had voice call prioritization capabilities (as a supplementary service) available since Release 6.

However, voice transmission functionalities are not the ones that are more popular for smart grid use. Data services are the ones used by most smart grid applications. There is no way to guarantee a service-level agreement (SLA) for data transport in most of the currently deployed public cellular networks. This is why, in order to protect the network planning and dimensioning of their networks, the different public cellular operators provide usage rules. Reference [19] showed the recommendations from network operators to vendors providing devices for M2M applications on their networks. These guidelines (common to other TSPs [20, 21]) have the objective of helping the M2M vendor "to deliver the best possible customer experience while maintaining the integrity of the [operator] network." The guidelines prevent against some device behaviors that are considered "aggressive" against the wireless network, that in some "extreme cases" can "cause temporary network failures or prolong the recovery time of network from such a failure.... Some of the behaviors that are considered normal in the wired network are not acceptable in the wireless network."

8.2.7 Security Aspects

Radio access, due to the nature of the medium, is available without a physical connection. This fact may appear as a weakness. If radio networks and systems are to be used in smart grids, they must be secure. Security in radio networks has evolved to cover all necessary features to take radio to the same security standards as any other network. This has happened both for private and public-oriented systems.

In the private radio domain, the first PMR systems evolved to have a very basic security to prevent the use of frequencies by unauthorized users with CTCSS (see EIA Standard RS-220-A, "Continuous Tone-Controlled Squelch Systems"). The mechanism consists in a certain sub-audio tone that is transmitted as signaling together with the useful signal and that makes the base station circuitry to repeat the signal only if the tone is recognized as the authorized one.

The digitalization of the systems introduced a more important security feature, the encryption of the end-to-end transmissions. Encryption is a security algorithm in the form of a mathematical procedure used to cipher data; the original information is encoded and requires a software key to be recovered. Some encryption techniques include [e.g., Digital Encryption Standard (DES) and Advanced Encryption Standard (AES)], but these are not the only ones, including a number of other proprietary ones in different commercial systems. Encryption techniques are also used to prevent unauthorized access to telecommunications networks through a secure authentication of users to prevent fraud, and to guarantee user data privacy.

Existing systems have progressively adapted these mechanisms. The first public cellular networks in voice-only 1G systems [e.g., advanced mobile phone service (AMPS)] were basically analog, with a very limited digital signaling [22]. Eavesdropping was straightforward, as the analog frequency modulation (FM) channel was unprotected, and terminal cloning and call setup signaling faking through simple recording were also easy. 2G systems (like GSM) evolved to try to overcome the limitations of the 1G systems. They came along with security for the mobile terminal based on a subscriber identity module (SIM) card. Authentication and over-the-air encryption were some of the major changes, thus protecting the access to the system [in collaboration with the authentication center (AUC), the home location register (HLR), and algorithms A3, A5, and A8, and keys in the AUC and the SIM card]. The life cycle of the user over the network (mobility management: registering process, mobile equipment location, call routing, and so forth), in all its different states contemplated security measurements to protect the identity of the user (e.g., temporal identities). The 3GPP TS 43.020 describes GSM security-related network functions.

In terms of data transmission, the application of data services in 2G systems is limited. These data services became popular with the 2.5G evolution (e.g., typically GPRS for GSM systems). Again, although there are improvements in the security for GPRS systems (security extending to the SGSN), this security is not integral. All in all, 2G systems do not have a security architecture as such and this is why 3G systems had to address this issue. For example, UMTS describes its security architecture in 3GPP TS 33.102 as the introduction of the 33 series where the rest of the aspects to include the core of the network in the system security are also covered. These 33 series also cover the evolution into the 4G (in the context of GSM, GPRS, and UMTS, this is LTE) in 3GPP TS 33.401.

8.2.8 The Machine-to-Machine (M2M) and Internet of Things (IoT) Paradigm

8.2.8.1 Introduction

Mobile-to-mobile (M2M) services and the IoT are referred to as groups of devices that connect to each other and/or central systems without, or with little, human intervention. These end devices include sensors and actuators. From a formal perspective, IoT lacks a formal definition (see [23]; the IEEE IoT initiative "to establish a baseline definition of IoT" includes a compilation of several definitions in [24]).

The M2M/IoT ecosystem is growing fast, as can be seen from the IoT/M2M Council [25] membership.

The reason to discuss M2M/IoT in a chapter devoted to radio for smart grids is its growing importance in the context of cellular public systems, and the direct relation between M2M/IoT and "smart" concepts (e.g., cities, appliances, and specifically meters and grids). Although radio is not the only technology directly related to M2M/IoT, it is probably the most popular. The ITU [26] sees in the IoT "an opportunity to capitalize on existing success stories, such as mobile and wireless [tele]communications."

M2M is usually understood within the IoT concept. In this sense, IoT might be understood as a generalization of M2M, collecting data from a broader range of end devices and processing it to extract useful information in a more complex and flexible way.

8.2.8.2 Architecture and Technologies

Both centralized and noncentralized architectures are proposed for IoT, depending on where the decisions are taken and how the communications happen. Centralized architectures are seen [27] as more appropriate for the purpose of IoT to minimize power consumption in end devices, keeping them simple and controlled (with some exceptions).

Although nearly all technologies can provide connectivity anywhere, it is not at any cost. Table 8.2 is a summary of the most relevant requirements [27, 28] of IoT networks for technology selection.

These requirements drive naturally to radio solutions as more appropriate than wireline technologies. Potential radio technologies for M2M/IoT applications can be classified into the following categories:

- Cellular networks: These existing networks offer a reasonable coverage for the population today. However, network coverage is not pervasive for static receivers, power consumption is high, and the evolution and replacement of the existing technologies is constant.

- Licensed spectrum radio networks: Licensed spectrum is an enabler of the control over the solutions deployed, as networks can be interference and congestion free if properly engineered. They typically allow for a higher power transmission than unlicensed ones, and thus coverage (and in-building penetration) is higher. However, licensed spectrum is neither easy nor quick to be found, and its cost is high (see [30]).

- Unlicensed spectrum radio networks: The advantages of these networks are evident (no need to obtain a spectrum license and no spectrum li-

Table 8.2
IoT Requirements

Requirement	Comment
Scalability	Billions of end-devices [29]
Cost	A few Euros/Dollars per each end-device (zero installation and configuration costs)
Replacement	Working life as long as the application itself
Power	Lifetimes exceeding 10 years on a battery (energy-harvesting options)
Coverage	Anywhere
Telecommunications reliability	Depending on the application type
Mobility	On-board vehicles and so forth
Security	Cybersecurity everywhere
Traffic	Small data packets and low-data rate (unicast, multicast, and broadcast)
Real-time capability	A must for certain applications; important in centralized architectures
Configuration	Auto-provisioning
Firmware upgrades	Over-the-air
Diagnostics and troubleshooting	No local operation

cense fee). However, unlicensed spectrum generally implies interference risk and limited power transmission (limitation of range).

8.2.8.3 Applications

The different M2M/IoT applications depend on the three main features that define the different radio technologies, as in Figure 8.9:

- Range: Local and personal area (Wi-Fi, ZigBee, and generic SRD standards), wide area (mainly cellular: 2G, 3G, and 4G), and global (satellite). This broad classification would affect, for example, coverage, location (indoor/outdoor), coverage, and mobility.
- Bandwidth: Narrowband and wideband. The classification combines, for example, data rate, volume of data, end-points duty cycle, and firmware upgrade.

  QoS: Low/medium/high. It affects, for example, security, criticality, sensitivity to delay, and error.

8.2.8.4 Standards

M2M/IoT references to standards can be found under either one of the two names and sometimes even under the name of machine-type communications (MTC) as in the 3GPP [32]. Furthermore, sometimes they can be found as part of previously existing telecommunication standards for which extensions are developed to cover new use cases (e.g., [33]).

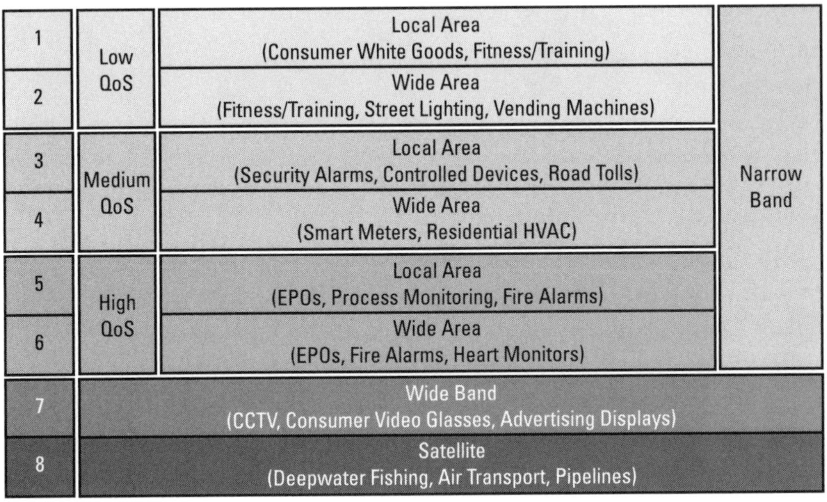

1	Low QoS	Local Area (Consumer White Goods, Fitness/Training)	Narrow Band
2		Wide Area (Fitness/Training, Street Lighting, Vending Machines)	
3	Medium QoS	Local Area (Security Alarms, Controlled Devices, Road Tolls)	
4		Wide Area (Smart Meters, Residential HVAC)	
5	High QoS	Local Area (EPOs, Process Monitoring, Fire Alarms)	
6		Wide Area (EPOs, Fire Alarms, Heart Monitors)	
7		Wide Band (CCTV, Consumer Video Glasses, Advertising Displays)	
8		Satellite (Deepwater Fishing, Air Transport, Pipelines)	

Figure 8.9 Matrix of applications and technologies, based on requirements [31].

Reference [24] compiled references to different standardization initiatives within various standardization bodies, among which we can see traditional telecommunications institutions such as IEEE, ETSI, and ITU, and others (IETF, NIST, OASIS and W3C), clearly showing the wide-ranging nature of M2M/IoT.

The IEEE has compiled a list of more than 100 IEEE standards related to IoT in [34]. The ETSI has been actively working on M2M since at least 2009 and has a technical committee (TC) developing standards for M2M communications. This TC has coordinated the work inside ETSI to contribute to the European mandates M/411 [35] and M/490 [36]. ETSI is part of the oneM2M [37]. The 3GPP discusses MTC in different reference technical documents and includes adaptation to M2M/IoT use in each of the incremental releases of their 3G and 4G radio technologies. ITU is active in IoT through the ITU-T Study Group 13 [38], which is involved in studies of IoT application support models, requirements for plug and play capabilities of IoT, semantic-related requirements and framework of IoT, architecture of IoT based on NGNe, and scenarios of implementing IoT in networks of developing countries.

8.3   Radio Technology Use Cases for the Smart Grid

This section includes two smart grid use cases that make extensive use of radio technologies and systems to enable both smart metering and smart grid applications.

8.3.1   Smart Metering Program in Great Britain

8.3.1.1   Introduction

The smart metering program is the answer to Directive 2009/72/EC of the European Parliament and of the Council of 13 July 2009 concerning common rules for the internal market in electricity, in the context of energy market regulation in Great Britain. The government's plans are to roll out an estimated quantity of 53 million smart electricity and gas meters to domestic and nondomestic properties in Great Britain by 2020.

In Great Britain the electricity meter is neither operated nor managed by the distribution (networks) companies; it is under the responsibility of the retailers. The level of deregulation in the market for Great Britain is high, and it is usual to find several retailers servicing customers in a certain area.

The DECC [39] is organizing the Smart Metering program in GB under the scheme represented in Figure 8.10 [40]. It is a customer-centric program where the rollout will be led by the suppliers with a framework of services provided by different companies. A shared smart metering national infrastructure provided by the data communications company (DCC) [41] is planned.

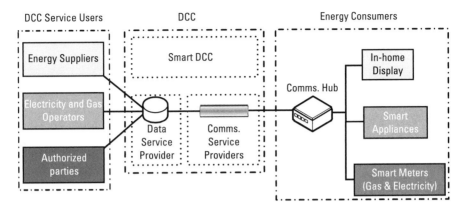

Figure 8.10   Great Britain smart metering system.

The system will cover electricity and gas meters and will offer in-home displays to complement the service offering the customer an easy access to their smart meter data. The DCC service users will be mainly the energy suppliers, but also the electricity and gas network operators that are entitled to use the data of the smart meters to better manage the grid that they are operating (notice that this is a way to guarantee that the information over the grid endpoints reverts to the network operators to use it for network operations). Any other entity may become a DCC service user, as long as they sign the Smart Energy Code (SEC), a multiparty agreement with the rights and obligations of the relevant parties involved in the end-to-end management of smart metering in Great Britain [42]. A DCC gateway connection is used to gain access to the data.

The telecommunication components in this framework lie within the responsibility of CSPs [with the smart metering wide area network (SMWAN)] and the energy consumers domain [smart metering home area network (SMHAN]. The communications hub is the element connecting the two networks, in the form of a gateway. This communications hub will have different implementations: one in the North and four different variants in the Center and South zones.

The communications hub has two functions: it allows the smart meters and in-home displays (and CADs) [43, 44] to communicate with each other over the HAN conformed by a ZigBee radio network; it also serves as the connection point to the WAN with different technologies depending on the service provider in the North, Center, and South zones.

The communications hub variants defined are [45]:

• CSP in the North will use one communications hub variant, provided by EDMI [46]. This will be an intimate communications hub [47]. There is a specific version, Fylingdales variant, which is intended to avoid the radar exclusion area of this zone (due to the frequency used).

- CSP in the Center and the South will be provided by Toshiba [48] and WNC [49] with four variants:
  - Cellular only, WNC;
  - Cellular only, Toshiba;
  - Cellular and mesh, cellular aerial port, Toshiba;
  - Cellular and mesh, two aerial ports, Toshiba.

Cellular specifications include [50]:

- WAN: Cellular communication module:
  - Dual-band GSM/GPRS/EDGE, class 10: 900/1,800 MHz (2G).
  - Dual-band HSPA: 900/2,100 MHz (3G).
  - Built-in antenna.
- HAN. IEEE 802.15.4 - 2.4 GHz (ZigBee): Built-in antenna.

Cellular-mesh specifications [50] include:

- WAN. Cellular communication module:
  - Single-band GSM/GPRS/EDGE: 900 MHz (2G).
  - Dual-band HSPA: 900/2,100 MHz (3G).
  - Built-in antennae + external SMA antenna port (automatic detection and switching).
- NAN. RF Mesh extension of WAN:
  - IEEE 802.15.4g RF Mesh networking at 869 MHz;
  - Built-in antenna.
- HAN. IEEE 802.15.4 - 2.4 GHz (ZigBee): built-in antenna.

A final mention should be made to the data component of this system, working on top of the telecommunication network. The application layer protocol of the smart meters will be DLMS/COSEM. CGI IT UK Limited [51] is the DSP that manages the movement of messages to and from smart meters.

8.3.1.2   Smart Metering Wide Area Network (SMWAN)

The DCC manages two contracts for the CSPs to put in place the smart metering WAN. Arqiva Limited [52] provides the network for the North region (Scotland and the north of England) using long-range radiocommunications.

Telefonica UK Limited's [53] network gives service in the Center (Midlands, East Anglia, and Wales) and South (south of England) regions using cellular radio communications plus mesh radio technology to extend the connectivity in hard-to-reach locations.

The typical operation of this smart metering system is as follows. Smart meters send their reading to the communications hubs in a preset schedule (e.g., every hour) and on another prearranged schedule the base station sends a message to the communications hub, that responds with the stored reading [54].

Long-Range Radio in UHF by Arqiva

Radio Parameters   Arqiva operates a licensed radio network in Great Britain, in ultrahigh frequency (UHF) band (412–414 MHz for uplink, and 422–424 MHz for downlink). The network equipment is provided by Sensus [55] and operates over a proprietary radio protocol.

The licensed band was acquired by Arqiva in a spectrum auction in 2006. The radio transmission conditions in the documentation of the tender [56] include the maximum transmission power allowed, and the out-of-band emission limits. The channels to be used are 12.5 kHz [57].

The maximum EIRP in the band 422–424 MHz is 47 dBm/25 kHz; the maximum EIRP of the base station is 56 dBm. The maximum ERP in the band 412–414 MHz is 10W.

The contractual documentation signed by Arqiva refers to the technical standards with which the solution complies [58]. Regarding spectrum use, reference is made to UK IR 2065 - UK Interface Requirement "Spectrum Access in the Bands 412 to 414 MHz Paired with 422 to 424 MHz" as issued by OFCOM [59], which include the details of the frequency allotment and its conditions.

System Architecture   The system deployed by Arqiva is based in Sensus's AMI system, called FlexNet. FlexNet is defined by the vendor in [60] as a single-tier system in the sense that the smart meters transmit directly to the base station without intermediate collectors or store-and-forward capabilities. The system, as described in the reference, transmits on the licensed band of 890–960 MHz spectrum with a patented RF modulation (this band is different to the one being used in Great Britain).

The architecture of FlexNet is built around the tower gateway base station (TGB), also called access nodes, that connects to the regional network interface (RNI) with primary and backup Ethernet connectivity to the network controller (NC). The connectivity of TGBs is achieved via DSL and VSAT technologies; alternatively, cellular 3G can also be used. The utility information platform collects the metering data from the NC to store it conveniently. The MDM is the middleware to present the information to the customer.

It is obvious that the system referred by the vendor needs to be adapted to work in Great Britain since the smart meter communicates through the communications hub, the frequency band allotment is different, and the architecture as a whole changes. The vendor defines the system as a solution for FANs and mentions it is based on point-to-point radio [61], meaning that the communication happens directly with the TGB. Thus, the wording might be misleading as the system is presented as a point-to-area or point-to-multipoint topology from base stations [60, 62]. The solution for the GB SMWAN is clearly not point-to-point in the way explained in this section.

The system has mechanisms for hard-to-reach locations [61], either through the "buddy mode" feature (which allows the communications unit to use a single hop to connect through its nearest neighbor, distances lower than 100m) or via a repeater solution connected to the TGB (e.g., rural areas where buddy mode is not an option). The system also has a macro-diversity gain as a result of the overlapping coverage of several cells, which is quantified at 3 dB and increases the resilience of the system.

Cellular and Mesh Radio Solution by Telefonica UK

Telefonica will be using its 2G/3G cellular network to support the SMWAN. This access network will be complemented with a low-power narrowband mesh radio technology supplied by Connode [63]. A software client will run in the residential communications hub as a complement to the access cellular network.

Cellular networks have poor coverage in some deployment scenarios such as cellars and some locations of specific building types. For this reason, the mesh radio technology (based on IEEE 802.15.4g) will be used as a complement.

Connode technology, according to [31], uses the 869.4–869.65 MHz band (250 kHz in one of the SRD bands with no specific use according to the ETSI EN 300 220-1), allowing for 10% duty cycle and up to 500 mW output power. Reference [64] claimed radio ranges of 0.4 to 5 km with low-energy consumption.

The vendor claims that the Connode 4 system is based on IP, and uses 6LoWPAN for IPv6 Connectivity, RPL, CoAP, DTLS, and AES-based transmission encryption.

From the architecture perspective, the existing cellular network will continue to use its traditional structure. Regarding the mesh radio system, according to [64], the Connode system consists of a number of meshed local radio networks called "micronets" and some elements called network control servers (head ends):

- The micronets consist of two types of nodes, this is, several standard elements and a master terminal. The master terminal acts as a gateway for hundreds of standard terminals and manages the communication with the network controller over any IP-based network connectivity.

- Each micronet is a flat and dynamic mesh network with multiple hops among terminals and a dynamic RPL routing mechanism.

8.3.1.3   Smart Metering HAN

ZigBee 2.4 GHz

The HAN in the smart metering program in Great Britain is the domain where the communications hubs, the smart meters and the CADs are connected. This connection is achieved via a ZigBee (IEEE 802.15.4-compliant; see Chapter 5) wireless connection in the 2.4 GHz band (operating within the 2,400–2,483.5 MHz harmonized frequency band, UK IR 2005 [7]). Once the CAD devices have been paired to their HAN, CADs will be able to access consumption and tariff information directly from their smart meter (electricity information every 10 seconds and gas information every 30 minutes [43]).

CADs are the way that businesses providing energy-related services or products may access energy data from the smart metering system, supporting the ZigBee SEP. The CAD could connect to the world outside the smart metering system (outside the DCC infrastructure) using non-DCC alternative communication means [43].

A special mention should be made about SEP. The SEP is a way to make application use the ZigBee technology. The smart metering Great Britain program prescribes ZigBee SEP v1.2 (described in the Great Britain Companion Specifications v0.8.1 [65]). ZigBee SEP v1.2 defines the device descriptions and standard practices for smart energy applications in residential and light commercial environment. Installation scenarios are varied, comprising from a single home to an entire apartment complex. The key application elements contained in the specification are metering, pricing (and scheduling), messaging and demand response, and load control. Commissioning and network managing are also contemplated in the specification. A key aspect of SEP is to guarantee the definition of standard interfaces and devices to allow interoperability of different ZigBee implementations.

HAN RF propagation trials suggest that 2.4 GHz radio will achieve approximately 70% coverage of homes in the Great Britain without using additional equipment, with ZigBee in a point-to-point or star type configuration [66]. The communications hub currently requires being able to support 24 links between smart metering devices on the HAN [67].

Reference [66] referred to a test in the University of Sheffield in 2011, with a system close to European regulations, with a system margin of 110 dBm (+8 dBm of transmission power and sensitivity of −102 dBm) and in an environment with three floors and a transmitter trying to reach all of them. The distances involved were lower than 20m and the results projected good enough communications in most of the locations. However, these results also showed the effect of floors and walls, as well as the different results with other values of system margin. These results are not surprising and should be used to correctly understand results in other geographical regions, where buildings are of a different composition and regulations allow for different power budgets. A further mention should be made to the importance of the hardware and software implementation of the platforms from the different vendors. Other test results are covered in [68]; they give more details on the number of errored packets depending on how far the different locations are (with the percentage reducing dramatically over a certain received signal level in distant locations), and how the different type of walls (structural versus partition, the first ones attenuate three times more) affect propagation.

Alternative HAN Communications

It has been recognized by the smart metering program that ZigBee in 2.4 GHz is not going to cover all the smart meters and CAD's possible locations, specially due to the difficult to reach scenarios (certain building types, e.g., large multiple dwelling apartment blocks) and specific installation circumstances of the smart meters.

The DCC has issued several public consultations [67] informing of the problems that have been found in the different assessments and field tests and to gather ideas on the solutions alternative to the use of ZigBee in 2.4 GHz. The two more important alternatives are ZigBee in 868 MHz and PLC solutions.

ZigBee in 868 MHz (under this denomination the government considers solutions in the 862–876 MHz and 915–921 MHz ranges, as covered by OFCOM's regulation) is anticipated in smart metering equipment technical specifications (SMETS2) as a complement to 2.4 GHz. Energy UK [69] is coordinating the metering industry in Great Britain to get the ZigBee Alliance to develop a suitable solution for the program in Great Britain. The 2.4 GHz and 868 MHz bands are estimated to provide HAN coverage in 95% of Great Britain premises (Red M report [70]); the communications hub would provide a dual 2.4 GHz and 868 MHz solution, the electricity meters would be prepared for 2.4 GHz, and gas smart meters could be 2.4 GHz, 868 MHz, or both. However, the details are not final.

PLC is one of the solutions proposed and studied for the 5% of the premises (approximately 1.5 million premises) that could be out of reach for ZigBee radio solutions. Although some participants to the public consultation have offered different variations of radio systems, it would seem reasonable that to solve the problems of radio coverage, other nonwireless alternatives could be used. PLC, both as narrowband option and as broadband (BPL) can be the alternative to bridge the gap not covered by radio. PLC as a collective (communal) and building-related infrastructure can be used specially in MDUs to create a HAN. Some studies, for example, [71], have been carried out in Great Britain buildings surveying the possibilities of narrowband PLC use in CENELEC B-, C-, and D-bands (as defined in EN 50065-1:2011), and for BPL from 2 to 30 MHz. However, gas smart meters would always be a challenge if a power socket is not available. This is why a mixed scenario where PLC and radio could be used seems a good option.

8.3.2 Public Radio Networks Used as a Multiservice Backbone

8.3.2.1 Introduction

The availability of a ubiquitous broadband wireless data communication service is highly attractive to connect locations where communications may be needed for multiservice connectivity purposes. This is the case with public mobile cellular networks that have evolved from the first generation (1G) to 2G, 3G, and 4G to support wireless data transmission.

During the period of deployment of these mobile networks, public commercial network users have been using existing fixed line commercial networks to connect private network end-points. Companies without private networks have used existing L2 services (FR, ISDN, and ATM first; progressively xDSL services) to create private networks. The 2G, 3G, and 4G public mobile cellular networks have progressively taken part in this connectivity evolution, as backup first, and eventually as a regular alternative when wireline services were not available or were too expensive.

8.3.2.2 Mechanisms for Private Network Creation over Public Networks

There are different options to configure private services over public telecommunication networks. These mechanisms are agnostic with respect to the nature of the network (i.e., wireline or wireless). Moreover, these mechanisms do not need any adaptation in the public network.

Wireless networks generally have lower availability than wireline counterparts; this is especially true in commercial public mobile networks. Moreover, the coverage of the mobile networks is not uniform across different TSPs. A single mobile TSP cannot offer unlimited coverage in the service area where a smart grid must be deployed, and access to a single mobile network means that a certain percentage of the service area will not be covered. Thus, it is advisable to consider wireless routers capable of hosting more than one SIM card (typically two). If one single radio exists in the router, it will not manage to move from one network to the other without connection termination (if higher availability is needed, a dual radio router might be considered). The preferred situation would be to have SIM cards that could be configured to dynamically connect to different TSPs' networks based on several criteria. The MVNO concept could also be considered to overcome SIM card property and management problems.

The connectivity mechanism proposed in this section is DMVPN, based on L3 IP connectivity. Some L2 options exist in the market.

DMVPN relies on two standard technologies, namely NHRP and multipoint GRE. DMVPN was first proposed by [72], but is a common, proven, and established solution with multiple implementations in the market.

The connectivity of end-points to a central element is achieved through hub-and-spoke architectures (Figure 8.11). However, these architectures do not scale well when site interconnection is needed, as many resources are used from the hub.

DMVPN solves the limitations of hub-and-spoke architectures:

- Multipoint GRE tunnels allow the establishment of a private network between the spokes and the central hub. This L3 network is totally unknown to the TSP, and inside the tunnels the customer will be able to make its private traffic progress. Apart from this, using IPSec (transport mode) increases the security as the traffic is encrypted. Although the TSP transports the traffic, it does not know the origin and destination of the private datagrams.

- Next Hop Resolution Protocol (NHRP), as per RFC 2332, creates a distributed mapping database of all the spoke tunnels to their public interface addresses. In this context, public refers to the L3 domain that is shared between the TSP and the customer, but it is based on private IP addresses agreed by both parties. Thus, what the TSP can see are

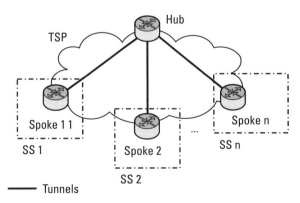

Figure 8.11   Hub-and-spoke architecture.

these NHRP IP addresses in the headers of the IP datagrams among the spokes and the hub, with the entire payload encrypted (including the GRE tunnel IP addresses).

In DMVPN the hub is a router, and all the spokes are configured with tunnels to the hub; there is no need to strictly define the tunnels in the hub because multipoint GRE and NHRP work dynamically creating the tunnel between them. Likewise, inside the tunnels, dynamic routing protocols will be possible as multicast and broadcast are supported. Thus, private addressing and services behind the spoke routers will be progressed.

8.3.2.3   Mechanisms to Increase Availability

Public networks are designed with reliability standards that are not usually adapted to the requirements of the services of the smart grid. This is why some extra mechanisms must be added to provide higher availability. This section shows how this can be achieved using the access to several mobile networks and a minimum of two interconnection delivery points between the third-party network and the private network of the utility, as shown in Figure 8.12.

Dual Interconnection Points Inside a Single Public Network

Network interconnection (between the TSP network and the customer's network) is critical in high availability configurations, where traffic cannot be lost

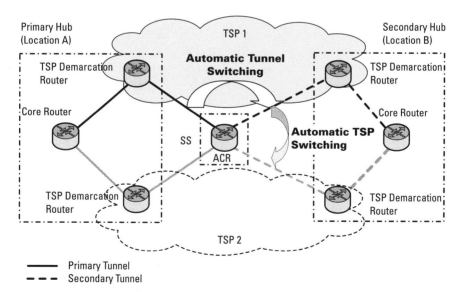

Figure 8.12   Optimal availability architecture for private networks over public commercial L3 networks.

if one single element fails. Thus, it is common to have redundant interconnection points at different places in the network.

DMVPN can be used in what it is known as dual hub DMVPN layout. The concept is based on two separate DMVPN environments, with each interconnection point (i.e., hub, two of which are in this example) connected to one DMVPN subnet; the spokes are connected to each of the DMVPN subnets. There are two L3 paths to connect the spokes with the core network and routing is needed to manage this redundant configuration. One option is to use different metrics (with dynamic routing) to access each of the interconnection points, balancing, or not, their use or simply to use one of them as a secondary backup connection, only activated if the primary path is not available.

Each end-point (spoke) has a tunnel to each interconnection point (hub). The activity on these tunnels can be checked through different mechanisms and protocols such as GRE keep-alives, IPSec DPDs (as per RFC 3706), or others. If tunnels are configured such that one is the backup of the other, and both of them are declared inactive by these mechanisms, it will mean that the network access is not available. As a consequence, only if the alternative network is available, the network will have a chance to provide service. In any other case, the only option is to loop trying to establish the tunnels until the network unavailability is solved.

Alternative Public Network Connectivity

Multiple public networks are available, and the way to get connectivity through any of them will be described in this section. This section considers access to public mobile networks.

If a public mobile network is to be accessed, a SIM card is needed. The typical situation is that public mobile operators offer their own SIM cards to have access to their network. As already mentioned, the availability of SIM cards not associated to specific TSPs would help managing the dependencies with public mobile operators. However, as this is not usually the case, multiple SIM cards must normally be used. It is common to have routers available that can hold and use alternatively two SIM cards, each with one TSP.

With routers that host one single radio, the first important aspect is the capability to check if the connectivity through the mobile network operator in use is available. Although the option to have two radios in a router (or two routers, each one with a single radio) is possible, it is not cost-effective. With one single radio, any of the mechanisms in the previous section can be used to check end-to-end network connectivity. However, apart from the possibilities mentioned, there are other radio-specific mechanisms that can be used in public mobile networks to understand if the radio access is available.

The specificities of different cellular network standards and the parameters configured in the different networks need to be considered when trying to use low-level RF parameter values to take decisions to connect to the alternative network. Examples of this can be found in 2.5 and 3G networks. In 2.5G GPRS networks, many decisions can be taken based on the RSSI value (commonly understood as coverage). In 3G networks [e.g., UMTS, RSCP, and E_c/N_0 (received energy per chip, divided by the power density in the band)] need to be considered, because the PHY is different. Timers (based on the configuration of the TSP network, which also defines its own policies to move users from, for example, 3G to 2.5G in case of network failure) need also be understood before a radio connection is considered lost.

In the previous context, it needs to be understood that even when the radio access might be working, there are other possible failure-prone points in the network where the service may be interrupted. This is the reason why it is necessary to check periodically at upper nonradio access layers that the traffic is properly progressing (at tunnel or IPSec level).

Once the connection with a TSP is lost, the connection mechanisms through the alternative one can be triggered. The algorithms to keep connection with one or the other TSP can be tuned to maximize availability or to optimize cost (if tariffs are different in the networks). This local intelligence is a fundamental issue in the routers, and the way they behave should be either fixed in the firmware versions or remotely tunable through configurable parameters.

References

[1] *Wireless Field Area Network Spectrum Assessment*, Palo Alto, CA: Electric Power Research Institute (EPRI), 2010. Accessed March 27, 2016. http://www.epri.com/abstracts/Pages/ProductAbstract.aspx?ProductId=000000000001022421.

[2] Sendin, A., *Fundamentos de los sistemas de comunicaciones moviles*, Madrid: McGraw-Hill, 2004.

[3] Young, W. R., "Advanced Mobile Phone Service: Introduction, Background, and Objectives," *The Bell System Technical Journal*, Vol. 58, No. 1, 1979, pp. 1–14.

[4] MacDonald, V. H., "Advanced Mobile Phone Service: The Cellular Concept," *The Bell System Technical Journal*, Vol. 58, No. 1, 1979, pp. 15–41.

[5] International Telecommunication Union - Radiocommunication (ITU-R), "ITU Terms and Definitions," 2010. Accessed March 27, 2016. http://www.itu.int/ITU-R/go/terminology-database.

[6] International Telecommunication Union - Radiocommunication (ITU-R), *Radio Regulations Edition 2012*, Geneva, 2012.

[7] UK Office of Communications (OFCOM), *IR 2005 – UK Radio Interface Requirement for Wideband Transmission Systems Operating in the 2.4 GHz ISM Band and Using Wide Band Modulation Techniques*, 2006. Accessed March 27, 2016. http://stakeholders.ofcom.org.uk/binaries/spectrum/spectrum-policy-area/spectrum-management/research-guidelines-tech-info/interface-requirements/uk2005.pdf.

[8] *ETSI Technology Leaflet: Short Range Devices*, Sophia Antipolis, France: European Telecommunications Standards Institute (ETSI), 2012. Accessed March 27, 2016. http://www.etsi.org/images/files/ETSITechnologyLeaflets/ShortRangeDevices.pdf.

[9] American Meteorological Society (AMS), "Glossary of Meteorology, 2nd Edition," 2012. Accessed March 27, 2016. http://glossary.ametsoc.org/wiki/Hydrometeor.

[10] National Electrical Manufacturers Association (NEMA), *A Brief Comparison of NEMA 250 and IEC 60529*, 2012.

[11] Utilities Telecommunications Council (UTC), *The Utility Spectrum Crisis: A Critical Need to Enable Smart Grids*, Washington, D.C.: 2009. Accessed March 27, 2016. http://www.utc.org/sites/default/files/public/UTC_Public_files/UTC_SpectrumCrisisrReport0109.pdf.

[12] Spencer, S., "Cellular Inadequate for Utilities, UTC Tells FCC," *Smart Grid Today*, 2011.

[13] Oldak, M., and B. Kilbourne, "U.S. National Broadband Plan RFI: Communications Requirements - Comments of Utilities Telecom Council," Washington, D.C.: Utilities Telecommunications Council (UTC), 2010. Accessed March 27, 2016. http://energy.gov/sites/prod/files/gcprod/documents/UtilitiesTelecom_Comments_CommsReqs.pdf.

[14] Hata, M., "Empirical Formula for Propagation Loss in Land Mobile Radio Services," *IEEE Transactions on Vehicular Technology*, Vol. 29, No. 3, 1980, pp. 317–325.

[15] ICT, *COST Action 231: Evolution of Land Mobile Radio (Including Personal Communication): Final Report*, 1999. Accessed March 27, 2016. http://grow.tecnico.ulisboa.pt/~grow.daemon/cost231/final_report.htm.

[16] Federal Communications Commission (FCC), "Wireless Priority Service (WPS)," 2015. Accessed March 27, 2016. https://transition.fcc.gov/pshs/emergency/wps.html.

[17] U.K. Government Cabinet Office, "Public Safety and Emergencies – Guidance: Resilient Communications," 2016. Accessed March 27, 2016. https://www.gov.uk/guidance/resilient-communications.

[18] Chambers, M. D., and D. H. Riley, "Implementing Wireless Priority Service for CDMA Networks," *Bell Labs Technical Journal*, Vol. 9, No. 2, 2004, pp. 23–36.

[19] Network Ready Lab, *AT&T Network Ready Device Development Guidelines*, AT&T, 2011. Accessed March 27, 2016. https://www.att.com/edo/en_US/pdf/NRDGuidelinesSeptember2011.pdf.

[20] Telstra, *Telstra Wireless Application Development Guidelines - Version 7 Issue 1*, 2014. Accessed March 27, 2016. http://www.telstra.com.au/business-enterprise/download/document/business-support-m2m-telstra-wireless-application-development-guidelines.pdf.

[21] Verizon Wireless, *4G LTE 'Open Access' Application Guidelines Version 2.0*, 2012. Accessed March 27, 2016. https://odi-device.verizonwireless.com/info/Open%20Development%20Device%20Docs/OpenAccessDocs/Guidelines_4GLTE_OpenAccessApplicationGuidelines.pdf.

[22] Køien, G. M., "Reflections on Evolving Large-Scale Security Architectures," *International Journal on Advances in Security*, Vol. 8, No. 1 & 2, 2015, pp. 60–78.

[23] IEEE Internet of Things, "About the IEEE Internet of Things (IoT) Initiative," 2016. Accessed March 27, 2016. http://iot.ieee.org/about.html.

[24] Minerva R., A. Biru, and D. Rotondi, *Towards a Definition of the Internet of Things (IoT). Revision 1*, IEEE IoT Initiative, 2015. Accessed March 27, 2016. http://iot.ieee.org/images/files/pdf/IEEE_IoT_Towards_Definition_Internet_of_Things_Revision1_27MAY15.pdf.

[25] IoT M2M Council (IMC). "About IMC." Accessed March 27, 2016. http://www.iotm2mcouncil.org/imcabout.

[26] ITU Strategy and Policy Unit (SPU), *ITU Internet Reports 2005: The Internet of Things*, Geneva: International Telecommunication Union (ITU), 2005.

[27] MacDonald, M., *The Internet of Things*, 2015. Accessed March 27, 2016. https://www.mottmac.com/download/file/127/6871/internet%20of%20things%20final.pdf.

[28] Alcatel Lucent, *The Case for M2M Device Management. How Network Operators Can Help Bring M2M to the Mass Market*, 2014. Accessed March 27, 2016. http://resources.alcatel-lucent.com/?cid=178525.

[29] Gartner, "Gartner Says 6.4 Billion Connected 'Things' Will Be in Use in 2016, Up 30 Percent from 2015," 2015. Accessed March 27, 2016. http://www.gartner.com/newsroom/id/3165317.

[30] Sawanobori, T. K., and R. Roche, *From Proposal to Deployment: The History of Spectrum Allocation Timelines*, CTIA The Wireless Association, 2015.

[31] *M2M application Characteristics and Their Implications For Spectrum – Final Report*, Aegis, 2014. Accessed March 27, 2016. http://stakeholders.ofcom.org.uk/binaries/research/technology-research/2014/M2M_FinalReportApril2014.pdf.

[32] Third Generation Partnership Project (3GPP), "About 3GPP Home," 2016. Accessed March 27, 2016. http://www.3gpp.org/about-3gpp/about-3gpp.

[33] "Machine to Machine (M2M) Communications Technical Report," IEEE 802.16 Broadband Wireless Access Working, November 2011. Accessed March 2016. http://ieee802.org/16/m2m/docs/80216p-10_0005.doc.

[34] IEEE Standards Association Innovation Spaces, "Internet of Things Related Standards," 2016. Accessed March 27, 2016. http://standards.ieee.org/innovate/iot/stds.html.

[35] European Commission, *Standardization Mandate to CEN, CENELEC and ETSI in the Field of Measuring Instruments for the Development of an Open Architecture for Utility Meters Involving Communication Protocols Enabling Interoperability*, Brussels, 2009.

[36] European Commission, *Standardization Mandate to European Standardisation Organisations (ESOs) to Support European Smart Grid Deployment*, Brussels, 2011.

[37] oneM2M, "oneM2M - Standards for M2M and the Internet of Things," 2016. Accessed March 27, 2016. http://www.onem2m.org/about-onem2m/why-onem2m.

[38] ITU-T Study Group SG13: Future Networks Including Cloud Computing, Mobile and Next-Generation Networks, Study Group 13 at a Glance," 2016. Accessed March 27, 2016. http://www.itu.int/en/ITU-T/about/groups/Pages/sg13.aspx.

[39]   UK Department of Energy & Climate Change (DECC), "About Us." Accessed March 27, 2016. https://www.gov.uk/government/organisations/department-of-energy-climate-change/about.

[40]   UK Department of Energy & Climate Change (DECC), *The Smart Metering System,* 2015. Accessed March 27, 2016. https://www.gov.uk/government/uploads/system/uploads/attachment_data/file/426135/Smart_Metering_System_leaflet.pdf.

[41]   Smart DCC, "About DCC." Accessed March 27, 2016. https://www.smartdcc.co.uk/about-dcc/.

[42]   The Smart Energy Code (SEC), "Becoming a Party to the SEC," 2013. Accessed March 27, 2016. https://www.smartenergycodecompany.co.uk/sec-parties/becoming-a-party-to-the-sec.

[43]   UK Department of Energy & Climate Change (DECC), *Smart Meters, Smart Data, Smart Growth,* 2015. Accessed March 27, 2016. https://www.gov.uk/government/uploads/system/uploads/attachment_data/file/397291/2903086_DECC_cad_leaflet.pdf.

[44]   British Electrotechnical and Allied Manufacturers' Association (BEAMA) – Consumer Access Device Working Group (CAD WG), *Consumer Access Devices: Applications for Data in the Consumer Home Area Network (C HAN) and Wider Market Considerations,* 2014. Accessed March 27, 2016. http://www.beama.org.uk/asset/EBA9BBB9-756E-48A0-B6C38EBFBC3EEEAB/.

[45]   Smart DCC, *Smart Metering Implementation Programme: Systems Integration Test Approach V5.0,* 2016. Accessed March 27, 2016. https://www.smartdcc.co.uk/media/352232/2016-02-12_system_integration_test_approach_v5.0_-_draft_vfinal.pdf.

[46]   EDMI Limited, "About EDMI," 2016. Accessed March 27, 2016. http://www.edmi-meters.com/About.aspx.

[47]   Smart DCC, *Intimate Communications Hub: Interface Specification V1.0,* 2014. Accessed March 27, 2016. https://www.smartdcc.co.uk/media/145112/intimate_communications_hub_interface_specifications_dcc_1.0_clean.pdf.

[48]   Toshiba, "Toshiba United Kingdom," 2016. Accessed March 27, 2016. http://www.toshiba.co.uk/.

[49]   Wistron NeWeb Corporation (WNC), "WNC Profile," 2016. Accessed March 27, 2016. http://www.wnc.com.tw/index.php?action=about&cid=1.

[50]   Meyer, R., *Datasheet - Communications Hub Cellular Only: SMIP Central/South Region v0.3 Draft,* Telefonica, 2014. Accessed March 27, 2016. https://www.smartdcc.co.uk/media/239468/datasheet_ch_central_south_tef_cellular_v0_3.pdf.

[51]   CGI, "Company Overview." Accessed March 27, 2016. http://www.cgi-group.co.uk/company-overview.

[52]   Arqiva. "About Us/At a Glance," 2016. Accessed March 27, 2016. http://www.arqiva.com/about-us/at-a-glance.html.

[53]   O2, "Telefónica UK Signs £1.5bn Smart Meter Deal," 2013. Accessed March 27, 2016. http://news.o2.co.uk/?press-release=telefonica-uk-signs-1-5bn-smart-meter-deal.

[54]　Waugh, M., *Arqiva's Smart Meter Network: Supporting Technical Justification for Site Reference LS016*, Altrincham, Greater Manchester: Arqiva, 2014. Accessed March 27, 2016. https://publicaccess.leeds.gov.uk/online-applications/files/1FDE3947A1FEE278CEF98207 CB64E99E/pdf/14_02309_DTM-SUPPORTING_TECHNICAL_INFORMATION-1000911.pdf.

[55]　Sensus. "Sensus Communication Technology Is Supporting Great Britain Smart Metering Project," 2013. Accessed March 27, 2016. http://sensus.com/great-britain-smart-metering.

[56]　UK Office of Communications (OFCOM), *Auction of Spectrum: 412-414 MHz Paired with 422-424 MHz – Information Memorandum*, 2006. Accessed March 27, 2016. http://stakeholders.ofcom.org.uk/binaries/spectrum/spectrum-awards/completed-awards/award412/im.pdf.

[57]　Joint Radio Company, *Response by JRC Ltd to Ofcom's Call for Information Promoting Investment and Innovation in the Internet of Things*, 2014. Accessed March 27, 2016. http://stakeholders.ofcom.org.uk/binaries/consultations/iot/responses/JRC.pdf.

[58]　UK Department of Energy & Climate Change (DECC), *Smart Meters Programme: Schedule 2.3 (Standards) (CSP North version) v.6.1*, 2013. Accessed March 27, 2016. https://www.smartdcc.co.uk/media/1317/5._Schedule_2.3_(Standards)_(CSP_Central_version)_[v.1].pdf.

[59]　UK Office of Communications (OFCOM), *IR 2065 – Spectrum Access in the Bands 412 to 414 MHz Paired with 422 to 424 MHz*, 2007. Accessed March 27, 2016. http://stakeholders.ofcom.org.uk/binaries/spectrum/spectrum-policy-area/spectrum-management/research-guidelines-tech-info/interface-requirements/IR2065May2007.pdf.

[60]　Sensus, *AMR-456-R2: FlexNet System Specifications*, 2012. Accessed March 27, 2016. https://distributor.sensus.com/documents/10157/32460/amr_456.pdf.

[61]　Sensus, *Answer to TEPCO Request for Comments on the Specification of Smart Meters and Our Basic Concept*, 2012. Accessed March 27, 2016. http://www.tepco.co.jp/corporateinfo/procure/rfc/repl/t_pdf/rep_059.pdf.

[62]　World Bank, *Applications of Advanced Metering Infrastructure in Electricity Distribution*, Washington, D.C.: World Bank, 2011. Accessed March 28, 2016. http://documents.worldbank.org/curated/en/2011/01/16357270/applications-advanced-metering-infrastructure-electricity-distribution.

[63]　Connode, "References: United Kingdom," 2016. Accessed March 28, 2016. http://www.connode.com/portfolio-item/united-kingdom/.

[64]　Westberg, M., "Internet – Soon Coming to a Meter Near You," *Metering International Magazine*, No. 3, 2012, pp. 48–49.

[65]　UK Department of Energy & Climate Change (DECC), *Smart Metering Implementation Programme: Smart Metering Equipment Technical Specifications Version 1.58*, 2014. Accessed March 27, 2016. https://www.gov.uk/government/uploads/system/uploads/attachment_data/file/381535/SMIP_E2E_SMETS2.pdf.

[66]   Egan, D., "ZigBee Propagation for Smart Metering Networks," *POWERGRID International*, Vol. 17, No. 12, 2012. Accessed March 28, 2016. http://www.elp.com/articles/powergrid_international/print/volume-17/issue-12/features/zigbee-propagation-for-smart-metering.html.

[67]   UK Department of Energy & Climate Change (DECC), *Consultation on Home Area Network (HAN) Solutions: Implementation of 868MHz and Alternative HAN solutions*, 2015. Accessed March 28, 2016. https://www.gov.uk/government/uploads/system/uploads/attachment_data/file/415578/HAN_Solutions_Consultation_March_2015_Final.pdf.

[68]   EDF Energy and Ember, *ZigBee Propagation Testing*, 2010. Accessed March 28, 2016. https://sites.google.com/site/smdghanwg/home/real-worldexperiences/EDF2010ZigBeePropagationTesting.pdf?attredirects=0&d=1.

[69]   Energy UK, "About Energy UK," 2015. Accessed March 28, 2016. http://www.energy-uk.org.uk/about-us.html.

[70]   Belloul, B., *Smart Meter RF Surveys - Final Report*, 2012. Accessed March 28, 2016. https://www.gov.uk/government/uploads/system/uploads/attachment_data/file/136124/smart-meters-rf-surveys-final-report.pdf.

[71]   Soufache, G., *Wired HAN Characterization Campaign: Test Report - Characterization of PLC Channels*, 2013. Accessed March 28, 2016. https://www.gov.uk/government/uploads/system/uploads/attachment_data/file/184798/TR_LAN12AF093-EnergyUK-HAN_PLC_Characterization_test_campaign_Ed03__2_.pdf.

[72]   Cisco Systems, *Dynamic Multipoint IPsec VPNs (Using Multipoint GRE/NHRP to Scale IPsec VPNs)*, 2006. Accessed March 25, 2016. http://www.cisco.com/c/en/us/support/docs/security-vpn/ipsec-negotiation-ike-protocols/41940-dmvpn.pdf.

9

Guidelines for the Implementation of a Telecommunication Network for the Smart Grid

Smart metering and the distribution grid modernization are a reality at many utilities on their way towards the smart grid. In most of these cases, the need to communicate with the smart meters goes with the opportunity to develop a telecommunication network that is either privately built for the purpose of this smart metering or takes advantage of existing public network services (a combination of the two is also frequent). This telecommunication network built for the purpose of the smart metering is suitable and likely supported on some of the existing grid assets and may also be used to provide additional services for other utility assets.

This chapter describes how a pragmatic smart grid evolution project can be developed. It starts from the need to provide smart metering services and takes advantage of this opportunity to develop a multiservice telecommunication network for the smart grid over the distribution grid.

9.1 The Conception Process of a Telecommunication Network for the Smart Grid

Utilities facing the smart grid evolution process ask themselves some common questions around telecommunications: the kind of infrastructure needed to support the telecommunications for the smart grid deployment, the features

267

that are going to be considered for this infrastructure, the objectives for the near future, and the effect on business operation are among them.

In this section, the reader will find the guidance for the definition of telecommunication networks for smart grid projects, considering both the strategic view and the impact on the electrical grid.

9.1.1 The Process

The development of a telecommunication network for a smart grid cannot be separated from the grid itself. As in any other business where information and communications technologies (ICTs) come to the aid of the business, the business objectives, and the vision on how these objectives need to be achieved are fundamental cornerstones of the process.

A smart grid deployment is a complex project that is not only made of technology, systems, infrastructure, and data, but also of organizations, resources, coordination, and legacy technologies. A group of high-level considerations are needed:

- Applications [e.g., advanced metering infrastructure (AMI), distribution automation (DA)] that are to be included in the smart grid;
- Changes in the operations as a result of the new applications;
- Organizational involvement in the smart grid evolution process
- Organizational skills within ICTs;
- Balance of internal and external human resources for the smart grid evolution (both the deployment, the operation, and the evolution afterwards);
- Balance of private networks and public commercial services.

In order to achieve a good transition from the traditional utility to a smart grid-enabled future-oriented utility, each company has to deeply understand how to organize the processes and resources to progressively integrate existing technologies within the daily operation of the company. The focus needs to be not on the excellence of each partial result (system or technology), but on the balance to achieve the overall objectives. All these aspects need to be complemented with a deep knowledge of the data and both the power system and telecommunication technology details, as they will be basic for the decisions on the smart grid development, deployment, operation, and maintenance [1]. Neither power engineers nor telecommunication engineers working independently will be able to develop this process.

9.1.2 The Adaptation of the Application Requirements

Once the applications that the smart grid needs to support are decided, the requirements that they impose on the telecommunication networks and services need to be derived. These requirements depend on the design of the application itself (e.g., different bandwidth or delay need for various grid assets) and on the way this application should perform to support the operation (e.g., smart metering with different data collection frequency needs for the groups of customers).

It is suboptimal to design applications independently of the telecommunication services that need to be deployed to support them. Thus, at the stage where the application functionalities and operation needs are to be defined, telecommunication networks and services constraints must be considered. Thus, the traditional and the new smart grid-related services covered in Chapter 4 need to be made explicit and translated into data requirements that are to be supported by the telecommunication networks in a technically and economically feasible way.

9.1.3 The Adaptation of the Telecommunication Network

The assessment of the current use of telecommunications in the utility should be accomplished to determine if there are synergies between the use of existing telecommunication networks and services, and what the utility intends to do in its evolution into the smart grid.

The assessment of the current situation should consider:

- Applications that are currently run and the characteristics of the telecommunication services supporting them (bandwidth, latency, and so forth).

- Single-service telecommunication networks or services that are only available to individual applications not within the context of a multiservice network vision.

- The network services inventory with the location where the services are delivered and the interfaces and protocols used.

- Private telecommunication networks available, with reference to the means used (optical fiber, wireless, and so forth), technologies (standard or not), and vendors. In all cases, the capacity (underuse and extramargin), age, and the obsolescence and support stage of each, needs to be understood.

- Public commercial telecommunication services currently used and currently available but not used by the utility (leased TDM/Ethernet point-to-point connectivity, L3 VPNs, wireless services). Price levels, available features, obsolescence stage, and coverage footprint are to be known.

- Telecommunication network management systems (NMSs) and the processes in place to monitor the quality of service (QoS) delivery in the network.

- Telecommunication network operation resources and skills with the different networks and technologies.

From a general perspective, utilities that already have a private network are in a better position to leverage their existing knowledge and resources. Utilities with a broad optical fiber network can possibly leverage this high-capacity asset to develop a core network; utilities with wireless technology-use tradition are often prepared to leverage the radio repeaters to extend their coverage in a quick and effective way. Utilities without NMSs tend to have the telecommunication knowledge spread within the different application-related departments and the management of the telecommunication network delegated in the utility control centers (UCCs). Multiservice telecommunication networks are the seed of effectively reaching the vision of a single multiservice telecommunication network where private and public resources are integrated to be centrally managed.

9.2   Design Guidelines

9.2.1   The Electrical Grid for the Support of the Telecommunications for the Smart Grid

Independently of the characteristics of the telecommunication networks in the different utilities, the grid assets themselves are the service end-points and as such need to be analyzed in their capacity to host telecommunication network elements. Moreover, the grid itself is an asset with an extremely important role in the creation of telecommunication networks, as it has been demonstrated by the use of rights of way, and networks on top or within the grid itself (e.g., optical fiber deployments on high voltage poles and ducts; radio base stations antennas and standard telecommunication cables supported by electricity grid poles; power line communications).

The deployment of the smart grid requires a transformation of the electrical distribution grid infrastructure. This infrastructure needs to closely consider the synergies of the grid and the telecommunication networks, especially relevant in underserved telecommunication areas, where the grid assets may be used to extend the telecommunication networks.

The grid and the telecommunication elements need to coexist in any case. The trend to reduce the available space in grid assets or the harsh conditions of certain grid sites are against a cost-effective introduction of telecommunication elements to provide service to them. It is obvious as well that not only the infrastructure but also the processes on the grid are affected by this integration of electricity and telecommunications, as operations need to understand the constraints introduced by the telecommunication devices. In a smart grid-ready future, any grid asset should be prepared to host or be serviced by telecommunication elements.

The grid situation is definitely different at the different voltage levels. In high voltage (HV) levels [e.g., in primary substations (PSs)], where control functions have been present for decades, electronics are common and so are telecommunications. However, medium voltage (MV) and low voltage (LV) distribution grid segments are not typically adapted to host telecommunications or electronics. In these cases, in the absence of suitable conditions, telecommunication elements must be adapted as described in the next sections, and costs will increase due to this circumstance.

9.2.1.1   Substations in MV

The variety of MV substation types is wide. This section matches the typical MV substation types and locations, with the available telecommunication means. The groups of substation types do not follow a strict classification that may cover all the evolution, electricity grid type and functional design, and the construction type of substations.

Different classifications of substations can be established. A broad classification follows population density of the area where they are usually installed. Urban areas usually present underground and in-building type of substations (e.g., ground level and basement). The density of the urban areas forces the integration of this type of premises that usually host several MV/LV transformers. The fact that the power lines are many times forced to be underground (a common European case), makes it unusual to find overhead MV power lines in these environments. Suburban areas share the in-building type with the urban areas, and include the shelter-type of substation and the pad-mounted transformers. In these cases, the MV power lines tend to be overhead, and are mixed with underground cable sections. Pole-mounted transformers are prevalent in rural areas, where a complete overhead power line distribution is usual.

There are, however, exceptions. Pole-mounted transformers might be found in urban areas (consider the overhead grid in the United States); in the outskirts of the cities, suburban type of constructions are found.

Different substation types allow different degrees of freedom for telecommunication network access. If the substation has not considered telecommunication services by design (which will be the case except for a few exceptions, e.g., ducts and cable entry when optical fiber cables are preplanned), an ex post selection of the best telecommunication alternative is needed (Figure 9.1):

- In-building type of secondary substations (SSs) is typically reachable by cable-based networks such as xDSL, hybrid fiber coaxial (HFC), and fiber to the home (FTTH). These networks have been extensively developed in buildings where the end-users live and can be used for smart grid services. It is true that these premises are not usually planned with ducts to connect with the general telecommunication cabling of the building, but if there is consensus with the property of the building, the work to connect with the telecommunications service provider (TSP) is usually small and affordable. Regarding the xDSL and HFC networks, as the telecommunication media is metallic, special care should be taken with the isolation of the grounding references in the substation and the TSP premises. This is not needed when the connection is made through optical fiber cables, where the telecommunication connectivity uses a dielectric media.

  Radio-based solutions can also be used. The antenna can be located inside the building or outside (walls visual impact and possible vandalism need to be minimized).

- Pole or pad-mounted transformers can be naturally approached with radio technologies, due to the overhead nature of the infrastructure. In these places, the installation of antennas to gain an appropriate coverage is possible. Outdoor cabinets are needed if a shelter is not available or

Figure 9.1 Secondary substations and telecommunication solutions.

has no room inside. The case of the poles is a special one as some country regulations make it possible that TSPs use electrical poles to develop their wireline telecommunication networks using this infrastructure (see, e.g., Title 47 Telecommunication of the U.S. Code of Federal Regulations (CFR), Section §1.1403 Duty to Provide Access; Modifications; Notice of Removal, Increase or Modification; Petition for Temporary Stay; and Cable Operator Notice).

- Surface shelter-type SSs can be also connected with radio technologies. In this case an indoor solution is usually enough and the position of the antenna becomes the most important issue. The antenna can be placed inside the shelter if the coverage is good enough, and should be installed outside (controlling the visual impact) if needed. There is usually a difference of 10 to 15 dB of better coverage outside the shelter.

- Underground SSs are not easily accessible with radio or by telecommunication cable-based networks. Radio signals have difficulties in penetrating underground, especially with the frequencies used by public mobile operators, and antennas cannot be made to improve coverage as no height or directivity can be achieved.

Power line communications (PLC) can also be mentioned as an appropriate telecommunication alternative for any SS type where PLC couplers can be installed on the MV feeders. PLC [typically broadband power line (BPL)] needs to be considered as a way to interconnect substations among them (PS to SS, and more often SS to SS) and extend the coverage to other substations where conventional telecommunication access (cable-based, radio, and so forth) cannot be granted (see Chapter 7). The typical point-to-point topology of underground networks, compared to the tree-like topology of overhead lines, is more controllable when a BPL network needs to be developed.

9.2.1.2 LV Grid

The LV grid is one of the most pervasive and yet an unknown asset for most of the distribution system operators (DSOs) worldwide. The LV grid flows from the SS to the end-customers, and gets continuity into the LV customer premises. The meter is always present as the point where the energy is measured and, from a telecommunication perspective, is another element part of the telecommunication network. Moreover, a detailed knowledge of the LV topology is needed to help smart metering telecommunication systems achieve their goals, integrating their telecommunication capability where it is needed (e.g., repetition points, both for wireless and PLC systems).

The LV grid does not attract much of the attention of the smart grid, except for smart metering-related applications. However, the LV voltage grid will be progressively important, as smart metering increases its resolution (more frequent meter reading, increased need to have all points of supply identified with feeder and phase for transformer balancing and so forth). Similar to the evolution of supervisory control and data acquisition (SCADA) systems and, in general, remote control from HV grid to MV, LV will see the advent of SCADA systems.

The LV grid assets of interest for the telecommunication network extension closer to the end-customers are (Figure 9.2):

- LV cabling: Cables in the LV grid can be found buried underground (with or without ducts) or placed overhead (hanging on walls or poles). Typically in three-phase systems, four conductors can be found, three for the phases and one for the neutral. LV cabling has a fundamental role to support PLC telecommunications. The cross-section of the cables and the splices in the cables do not affect the transfer of energy power supply, but can be relevant for PLC telecommunication signal distribution, especially in the higher frequencies associated to BPL. Loads connected

Figure 9.2   LV grid assets.

to the grid also affect the behavior of PLC systems, specifically if they introduce noise in the grid in the frequencies where PLC works. Intermediate injection points in the LV grid can be used for signal repetition.

- Poles: Overhead LV lines can be laid on poles or hanging on walls of different buildings when this is possible. Poles can be made of concrete, wood, or metal and they are structures where PLC repetition points can be implemented, and wireless signal repeaters can be installed. The same is possible on walls, where any telecommunication equipment hosting-cabinet could be fixed. The strong point of these places as telecommunication network support elements is that power supply is available.

- Cabinets: Along the LV line path, different types of cabinets can be found. These cabinets perform different functions [cable distribution, demarcation, and security (fuse boxes, protection, and measurement enclosures)]. The room available in these cabinets is usually limited, but can still be used to host telecommunication devices.

- Built-in electricity meters: In an effort to avoid the installation of meters inside homes, detached and semidetached houses many times present built-in wall enclosures for meters so that they can be read without entering the property of the customer. These places are an alternative for installation of telecommunication network elements if their size is small.

- Meter rooms: These are the rooms (usually in the basement or at the top of buildings) where the utility meters are located. Where they exist, typically regulation allows the installation of auxiliary telecommunication devices to perform electricity metering and control activities.

9.2.2 Technical Specifications for Network Equipment and Public Services in the Smart Grid

Even if telecommunication requirements for devices and services are fundamental, some of the most challenging requirements from the perspective of either a product vendor or a service provider are related to the adaptation to the grid conditions. As a consequence, the market does not offer, in general, telecommunication solutions (equipment and services) natively designed for HV and MV grid scenarios.

Regarding equipment, although the distribution grid is usually considered an industrial environment, special grid-related aspects such as size, electric isolation, temperature range are new, and so forth.

With regard to TSP services, the service conditions normally delivered by public networks do not in general cope with some of the requirements of the smart grid (e.g., demarcation devices environmental or access conditions, mission-critical aspects).

9.2.2.1 Nonfunctional Technical Specifications for Network Equipment in Secondary Substations

Nonfunctional technical specifications refer to all the characteristics of equipment that are not related to telecommunication functions. There is abundant literature on specific requirements for devices to be installed at substations and similar locations, particularly in the International Electrotechnical Commission (IEC) and the Institute of Electrical and Electronics Engineers (IEEE) standards.

A well-known source is IEC TC 57 "Power Systems Management and Associated Information Exchange." Two series of standards set requirements for smart grid-related equipment: IEC 60870 "Telecontrol Equipment and Systems" and IEC 61850 "Communication Networks and Systems for Power Utility Automation." The IEC 60255 series (originated in IEC TC 95) gives requirements specifically for protection devices.

In IEC 60870, Part 2 "Operating Conditions" comprises two relevant international standards:

- IEC 60870-2-1:1995 focuses on electromagnetic compatibility.
- IEC 60870-2-2:1996 focuses on environmental conditions (climatic, mechanical, and other nonelectrical influences) and partially supersedes IEC 60870-2-1:1995.

In IEC 61850, Part 3 "General Requirements" focuses on three different areas:

- From the functional perspective, it covers both automation and the associated telecommunications.
- From the application perspective, it gives requirements for individual intelligent electronic device (IEDs) and complete "systems" [substation automation system (SASs), including IEDs and the telecommunication network infrastructure].
- From the environment perspective, it is intended for substations and power plants.

Although there is a difference of 15 to 20 years between both sources (IEC 61850-3 and IEC 60870-2), many provisions in IEC 60870-2 are still valid. The scope of the IEC 60870 series is telecontrol equipment and systems for monitoring and control. In this sense, it is not necessarily constrained to substation environment, hence its interest for a case study that includes elements to be located at geographically widespread locations. Thus, although the scope of both standards is different, both will be considered for the purpose of this section.

From the IEEE perspective, WG C2 is responsible for standard IEEE 1613 that specifies "standard service conditions, standard ratings, environmental performance requirements and testing requirements for communications networking devices installed in electric power substations." IEEE 1613-2009, IEEE 1613a-2011, and IEEE 1613.1-2013 have broadened the scope to include devices installed in all electric power facilities, not just substations, and include the telecommunication modules installed in smart meters. This applies to devices used for DA and distributed generation (DG). It also adds specific requirements for radio frequency (RF), PLC, or Ethernet-cable devices.

Electrical Requirements

Active devices that are to be part of a smart grid deployment usually have to respect specific constraints in terms of voltage levels and tolerance, power supply redundancy, battery lifetime, and so forth. These constraints are normally specific to the type of equipment and its location.

Both IEC 60870-2-1:1995 and IEC 61850-3:2013 give some general guidelines to be considered for equipment specification. For example, for voltage levels, in cases where the device is to be supplied from alternating current (ac) public mains, 120/208V or 230/400V values have to be supported (although some other values may have to be specified depending on the region).

Tolerance on the ac input is also important so devices can work normally with voltage ratings that are above or below nominal. IEC 61850-3:2013 recommends that the operating range be 80% to 110% of the rated voltage. IEEE 1613.1-2013 only requires 85% to 110% of rated voltage at rated frequency. IEC 60870-2-1:1995 actually classifies equipment by the nominal voltage tolerance it supports: classes AC1, AC2, AC3, and ACx are described. Class AC3 means the tolerance range spans from −20% to +15% of the rated voltage, which represents the highest tolerance and the one recommended for smart grid equipment. In the typical European case of 230V at 50 Hz, this means a wide operational voltage range from 184V to 265V. IEC 60870-2-1:1995 additionally defines different classes for ac frequency tolerances (up to ±5% of the nominal frequency).

Ac inputs to be connected to 230/400V systems will also show compliance to IEC 61000-3-2:2014 with regard to harmonic current limits to be injected into the mains and IEC 61000-3-3:2013, which gives limits for short-term and long-term flicker values and voltage changes.

In the case of direct current (dc) inputs, it is also important to specify the voltage ratings at which the equipment will operate. Traditionally, one or more of the following values are used for smart grid equipment: 12V, 24V, 48V, 60V, 110V, 125V, 220V, or 250V. Again, IEC 60870-2-1:1995 defines class DC3 for voltage tolerance (−20% to +15% of the rated voltage) that is the recommended range for most smart grid scenarios (similar to dc ratings in Section 4.1 of IEEE 1613.1-2013).

EMC Requirements

Regarding conducted or radiated emission limits with which smart grid devices have to comply, CISPR 22:2008 is usually considered the reference. It sets conducted limits, in the frequency range from 150 kHz to 30 MHz, for power supply ports, and telecommunication ports in the case of telecommunication devices. Additionally, it establishes radiated emission limits above 30 MHz. These limits do not apply in the case the device is a radio transceiver.

From the perspective of immunity, smart grid devices in power plant and substation environments have to be designed to withstand the different kinds of conducted and radiated electromagnetic disturbances, both continuous and transient, that occur in electric power systems. A majority of EMC immunity requirements stems from tests described in the IEC 61000-4 series.

ESD immunity testing according to IEC 61000-4-2:2008 is needed for every device to be installed in smart grid scenarios, where the highest severity test level is recommended (i.e., 8 kV contact discharge). This is, according to IEC 60870-2-1:1995, the required level for units installed in uncontrolled areas. It is also the recommended test level for electric power facilities physically located within a boundary or fenced area—such as a generating station or a transmission or distribution substation—per IEEE 1613.1-2013. A test level of 6 kV for contact discharge would be acceptable only for equipment to be installed in dedicated rooms with humidity control, in line with IEC 61850-3:2013.

Immunity to radiated RF electromagnetic field (EMF) shall be tested according to IEC 61000-4-3 and applied to all kind of products, as exposure to RF fields cannot be practically avoided. A test level of 10 V/m for the 80–3,000-MHz range is recommended both by IEC 60870-2-1:1995 and IEC 61850-3:2013; this is generally the protection required for industrial environments. This test guarantees that, for example, smart phones and laptops in proximity to deployed equipment will not cause malfunctioning. A more complex text is required in Section 7 of IEEE 1613.1-2013 (based on IEEE C37.90.2-2004 for relays, and specifically designed for the operating practices in the United States), covering frequencies up to 5.8 GHz.

The conducted version of the RF EMF immunity test is described in IEC 61000-4-6:2013. Frequencies from 150 kHz to 80 MHz are tested and the accepted test level in the utility industry corresponds to 10V of open-circuit test level, expressed in root mean square (rms) value of unmodulated disturbing signal.

The electrical fast transient/burst immunity test according to IEC 61000-4-4:2012 should be applied to any device that is connected to mains or has cables (either signal or control) in proximity to mains. The latter case may be generally applied to any equipment present in the smart grid. The test is intended to demonstrate the immunity of equipment to transient disturbances such as those originating from switching transients (e.g., interruption of inductive loads, relay contact bounces). The recommended test level is 4 kV peak for power or protective ground ports, and 2 kV peak for control and signal ports. This corresponds to severe industrial environment, which is to be found in outdoor electric premises and power plants or substations. For reference, IEEE 1613-2009 and IEEE 1613.1-2013 define a different approach with similar test levels.

Products that are connected to networks leaving the building or mains in general (e.g., ac-powered equipment) will also be tested for surge immunity as described in IEC 61000-4-5:2014. This procedure tests against overvoltages from switching and lightning transients or faults. Test levels 3 (open-circuit test voltage of 1 kV line-to-line, 2 kV line-to-ground) or 4 (double the values) are recommended, with a 1.2/50 μs waveform for power ports and 10/700 μs waveform for telecommunication ports connected to long lines (e.g., xDSL).

In the case of ac power ports directly connected to mains, immunity to voltage dips (caused by faults in the network, e.g., short circuits), in installations or by sudden large changes of load, is tested according to IEC 61000-4-11:2004. IEC 61850-3:2013 defines a voltage dip of 30% for 1 period and a voltage dip of 60% for 50 periods. The equivalent for dc power ports is the test specified in IEC 61000-4-29:2000, for which a voltage dip of 30% for 0.1 second and a voltage dip of 60% for 0.1 second are defined.

Similar to voltage dips, voltage interruption tests are defined in the same standards. For ac power ports, IEC 61850-3:2013 recommends an interruption for 5 cycle periods and another one for 50 cycle periods. In the case of dc power ports, the test recommendation is for an interruption of 0.05 second.

Regarding mechanical requirements, they depend largely on the type of equipment, but in general, it shall be specified if the equipment will be 19-inch rack-mounted or if it should support outdoor installation. Dimensions of the devices should be limited if they are going to be installed in constrained areas. Weight may be also a concern, and fall tests could be performed if preassembled cabinets and devices are going to be transported. IK code according to IEC 62262 is used to specify the degrees of protection provided by enclosures for electrical equipment against external mechanical impacts.

The IP code shall be provided according to IEC 60529. This is a standard that indicates the degrees of protection provided by enclosures. Protection is tested against ingress of solid foreign objects and against ingress of water with harmful effects.

Environmental Requirements

Environmental engineering is a broad concept that encompasses both climatic and mechanical (vibration, shock, seismic) tests. The ETSI produced a series of standards, ETSI 300 019 that are a good guidance for the specification of smart grid equipment mainly serving telecommunication purposes. Most tests are based on the classical IEC 60068-2 series "Environmental Testing - Part 2: Tests."

Specifically, every piece of equipment will have to be specified for:

- Storage: According to tests in ETSI EN 300 019-2-1. For example, if storage is to be done at weather-protected, not temperature-controlled, locations, the standard defines tests for low and high air temperature according to IEC 60068-2-1 and IEC 60068-2-2, high relative humidity as per IEC 60068-2-78, condensation as per IEC 60068-2-30 and vibration tests, sinusoidal according to IEC 60068-2-6 and random according to IEC 60068-2-64.

- Transportation: According to tests in ETSI EN 300 019-2-2. For example, if equipment transport is to be done by public transportation, the standard defines tests for low and high unventilated air temperature according to IEC 60068-2-1 and IEC 60068-2-2, air temperature change according to IEC 60068-2-14, relative humidity slow temperature change as per IEC 60068-2-78 and rapid temperature change according to IEC 60068-2-30, rain intensity as per IEC 60068-2-18, random vibration test according to IEC 60068-2-64, shock tests according to IEC 60068-2-27 and free fall test according to IEC 60068-2-31.

- In-use: One or more classes have to be selected among those in ETSI EN 300 019-2-3 to 2-8.

Each of these standards describes the specific tests depending on the environmental class selected. For example, for a typical case in smart grids, which is stationary use at weather-protected, not temperature-controlled, locations, tests are defined for low air temperature according to IEC 60068-2-1, high air temperature according to IEC 60068-2-2, air temperature change according to IEC 60068-2-14, high relative humidity according to IEC 60068-2-78, condensation according to IEC 60068-2-30, vibration tests according to IEC 60068-2-6 and IEC 60068-2-64 and finally shock tests as per IEC 60068-2-27.

IEEE 1613.1-2013 defines a simpler set of requirements: one of four possible ranges of operational (i.e., in-use) temperature shall be declared by the manufacturer along with one of three possible ranges for nonoperational temperature (i.e., transport, storage, and installation); relative humidity parameters along with vibration and shock severity classes taken from IEEE C37.1-2007.

9.2.2.2 Functional Technical Specifications for Telecommunication Equipment in Secondary Substations

Apart from the nonfunctional requirements defined above, it is obviously necessary to define technical specifications describing the functional requirements of the telecommunication devices, in order to guarantee its performance within the network architecture and mitigate potential interoperability troubles. Thus, routers, switches, and wireless and PLC devices must implement all the protocols and functionalities considered mandatory for its smart grid deployment.

These functional specifications start with the description of a first level of minimum mandatory aspects that the individual devices need to comply with. The following list can be used as a reference:

- Interfaces: The number and characteristics of the different device interfaces. They include aggregate and tributary interfaces, together with any antenna port that could be needed. The physical and logical feature compliance must be defined for each of the interfaces.

- Technology-specific functional aspects: They need to be defined with reference to standards, avoiding proprietary definitions as much as possible. If any proprietary feature is requested, the level of detail of its description should make any vendor capable of implementing it.

- User interfaces for equipment control: These interfaces should allow the configuration and control of the device both locally and remotely [e.g., command line interface (CLI) and graphical user interface (GUI)].

- Monitoring capabilities: The devices should be monitored during its operation to check their performance.

- Troubleshooting resources: The means to be able to locate and solve any device malfunctioning: troubleshooting (ping, trace route, and so forth), logs, and statistics.

As examples of these aspects, the minimum number of ports (electrical and optical) should be specified for Ethernet switches. In 2G/3G/4G radio routers, it is critical to determine if single or multiple subscriber identity module (SIM) cards will be used.

A second set of requirements should provide a detailed and exhaustive list of functional requirements for every device in a network context. As a continuation of the example of the switch, it is important to indicate the protocols and standards that it should be able to manage (e.g., IEEE 802.1D-2004, IEEE 802.1Q-2014, RADIUS, NTP, and DHCP).

Finally, the functional specifications have to be complemented with a set of tests that needs to be performed in order to verify the proper operation and interoperability of devices. Table 9.1 includes an example of tests for double SIM-card 2G/3G/4G routers.

Table 9.1
Example of Tests over a Double SIM-Card 2G/3G/4G Radio Router

Test Groups	Description
WAN. SIM card performance	Verify basic operation of main SIM: connection establishment and tunneling
WAN convergence	Verify convergence time between SIMs (main SIM and secondary SIM) in every scenario (switchover cases)
Routing protocols	Verify network routes, checking if they are correctly advertised with the selected routing protocols
IPsec and NHRP	Verify IPsec checking the proper operation in the router (phase 1, phase 2, security associations management) and NHRP-based tunnel monitoring mechanisms
LAN functionality	Verify LAN/VLAN setups to enable/disable traffic flow among devices and services connected to the same LAN
L2 functionality	Verify multicast, STP/RSTP, port mirroring
Device management	Verify access to device management via local/remote connections; SNMP compliance
Device access security	Verify the controlled access to the router, with each of the implemented protocols and services (RADIUS, IEEE, 802.1X, SSL, LDAP, and so forth)
Device monitoring	Verify the correct functional behavior of the different alarms, checking log files and the device LEDs
Firmware upgrade	Verify the correct upgrade of the firmware both locally and remotely
Log collection	Verify the collection of events and their correct classification in the internal logs
MAC filtering	Verify the mechanisms to limit network access to selected devices
Virtual Router Redundancy Protocol (VRRP)	Verify VRRP behavior in every high availability scenario
QoS functional testing	Verify QoS mechanisms to managed congestion and to ensure high priority traffic is handled correctly
Others	Verify NTP synchronization, DHCP, configuration files

9.2.2.3 Technical Specifications for Public Telecommunication Network Services Delivered at Secondary Substations

The definition of the technical requirements provided by TSP's contracted services for smart grids requires specifying not only technical requirements, but the rest of activities that would typically be performed by the utility if the services were provided by a private network.

From the point of view of the service provision, the coverage area is required. The definition of the service provision and network repair aspects are also required. From the technical requirements perspective, the connectivity, the management, the maintenance and the monitoring of the service need to be defined, together with some other characteristics such as availability, security, and the data tariff model on the network. Technology future changes must be known, as they would imply changes in the network access terminals.

Coverage area is an important part of the service. In the case of wireline technologies, it requires that the TSP exchange is prepared and activated for the service (xDSL, HFC, FTTH) and that the cables reach the destination where the service is needed. Distance aspects over existing cables must be taken into account (not all locations are within the distance range to offer certain bandwidths) and the existence of cables or the need to deploy new ones must be considered. In the case of radio technologies, the frequency bands allocated to the different radio technologies used by the TSP (2G, 3G, 4G) should be known as they will limit penetration in buildings. Geographical coverage must be broad if we want to cover the range of SSs in a typical smart grid deployment, as they are typically close to populated areas (typical surface coverage of wireless service TSPs is lower than population coverage). Radio coverage data is generally provided at ground level, and this does not always correspond with the available location of the antenna in SSs (e.g., underground).

Regarding TSP network availability, which is especially relevant when the service provided is mission-critical, network exchanges and radio repeaters should have battery autonomy in accordance with the required service levels. This is a classical service problem in electricity control applications of public radio networks, because in the case of a power outage affecting a radio repeater, if the autonomous power supply is not well dimensioned, the public telecommunication network will not be available to remotely manage the electrical network element and help to recover the grid fault.

The TSP's core network has also a role in service availability, and with the appropriate internal system architecture redundancy, must be transparent to the service. The redundancy should also be stretched to the network interconnection sites between the TSPs and the utility's network. The mechanisms to be established by the TSP will guarantee a permanent service delivery. If the TSP service redundancy cannot be improved, utilities should consider duplicated service delivery via other alternative TSPs.

Security needs to be considered in a public network connection. In this context, the use of virtual private networks (VPNs) is recommended to guarantee isolation among different customers using the same carrier network. In public mobile networks, there must be an appropriate registration method and a strict management of SIM profiles that allows connecting new smart grid premises. The use of machine-to-machine (M2M) SIM cards provides management tools for the utility to activate or deactivate connections, control inventory, monitor alarms, and so forth.

Bandwidth capacity and data flows prioritization must be guaranteed by the TSPs' network and system mechanisms. TSPs must provide details of how they implement it. These mechanisms will be engineered in such a way that will ideally guaranty that under no circumstance high-priority traffic (e.g., DA) can be dropped or delayed in favor of lower-priority traffic (e.g., AMI). Bandwidth capacity needs to be dimensioned to cope with all the traffic offered to the network. The aspects mentioned before result in a certain performance level that should be established in the specifications. They will be generally defined as some minimum threshold limits required for each of the services provided by TSPs: throughput, latency, jitter, packet loss, bit error rate, and so forth. These requirements will be part of a service level agreement (SLA) in the form of a contractual commitment from the TSP, so that the services can support the smart grid deployment, operation, and maintenance. The contract should also include mechanisms to measure and control the performance parameters granularly (circuit level), ideally in real time but commonly with a certain periodicity (weekly, monthly). A network audit to check bandwidth availability at the interconnection points, prioritization mechanisms, and protection and backup is needed as well.

9.2.3 Criteria for the Evolution of the Electricity Grid Infrastructure

The adoption of telecommunications in the grid should not be based on the adaptation to a certain technology today, but on the adaptation of the grid infrastructure to support the evolution of the telecommunication network.

This infrastructure adaptation is related to the reserve of room for telecommunication solutions to allow the use of electricity grid assets by telecommunications, the adaptation of electrical sites to telecommunication environmental conditions, and the arrangement of the infrastructure for telecommunication services access:

- Room must be reserved in substations to host electronic devices. This space should be environmentally ready to host typical industrial or commercial-type telecommunication. Room must also be present for power backup elements.

- Access to telecommunication networks should be present by default at substations. If the network is a wireline one, there should be ready-to-use ducts to connect the room reserved for the electronic equipment with the outdoor surrounding. If the network is wireless, there should be a provision of the antenna space, in such a way that it is easily accessible for technicians and not visible for the general public; in the case of underground substations this antenna installation could be challenging and require external poles for this purpose and the ducts to hold the cables complementary to the radiating system.

- Power lines, when buried, should be ducted and extra ducts should be laid to support telecommunication connections among substations. This interconnection is typically achieved with optical fiber cables (although traditionally copper-pair cables have been used for this connection). A minimum of two connections via separate physical routes must be supported. This power line burying is typical of HV and MV grid, but will increasingly happen in LV grid, where ducts could also be helpful to reach building meter rooms to improve the reachability of these sites close to the customers.

- Power lines that are not underground are usually suspended on poles or walls. The rights of way can be used (and are in fact used so by many utilities and TSPs, particularly cable network operators) to support optical fiber cables. The classical cables, exclusive of utilities, are the optical ground wire (OPGW) (see IEC 60794-4-10, IEEE 1138-2009, and IEEE 1591.1-2012) and the optical phase power cable (OPPC) (IEC 60794-4, that also covers OPGW and ADSS); there are other types commonly used such as the all-dielectric self-support (ADSS) (see IEC 60794-4-20 and IEEE 1222-2011). Of late, FTTx cable designs have opened the way to new flexible urban deployments that can follow the track of electric cables on walls and noncontinuous surfaces.

- MV switchgears in substations should be prepared for BPL injection. This involves both the connectors (that should be standard size and equal all through the utility) and the couplers that could be preinstalled in the switchgears (to avoid their de-energization to install couplers).

- LV panels should have appropriate connection points and room to install any electrical device to couple PLC signals, as well as any inductive device to get current information per feeder (or phase in the feeder), to obtain power information.

- Electricity meters, and the rooms or enclosures in which they are installed, should have standard interfaces to connect any telecommunication terminal (PLC, wireless, and so forth) for its connectivity and control.

- For grid operation adaptation, some operational grid processes will now be improved by the availability of telecommunications; some others will be constrained by the existence of telecommunication elements (e.g., a power line fault could interrupt a PLC link working on it).

9.3   Telecommunication Network Deployment for the Smart Grid Use Case

This section will describe a use case [2] to implement the telecommunication network needed to deploy a smart grid over the distribution grid to support an AMI and DA services.

9.3.1   Secondary Substations and Services

In the context of tens of thousands of SSs, with an average of customers per substation of more than 100, not all the substations have the same relevance. For example, the density of customers is higher in urban and suburban areas than in rural areas; a problem in a urban substation affects more customers than in a rural one.

From a smart grid services need, the SSs of this use case will be categorized as:

- Basic secondary substation (BASS): Smart metering services will be provided to the meters connected through this substation. AMI services will be remote reading, remote connect and disconnect, and meter connected-phase identification. No other added value will be provided to the substation itself.

- LV monitored secondary substation (LMSS): This category will have the services provided to BASS, and will include LV supervision, a service provided through a totalizing meter to measure the total power delivered. It could include LV meter feeder identification.

- MV monitored secondary substation (MMSS): This category will have the services provided to LMSS, and will include MV grid monitoring functions to retrieve phase voltages, currents, active and reactive power, wave quality information, and events reporting.

- Automated secondary substation (AUSS): This category will have the services provided to LMSS and will include MV remote control of switchgears and automation of recovery processes.

- Advanced security secondary substation (ASSS): This type is not another category, but an attribute that can be assigned to any of the other categories. The reason is that not all the substations need the same level of physical protection, with theft being an important driver of this attribute.

These service categories are related to the applications to be deployed. AMI is related to all the customers' meter base, and no SS can be excluded from this requirement (especially as it is many times related to a mandate imposed over utilities). DA, on the contrary, is not always deployed at every substation, as depending on the grid topology and on the SAIDI or SAIFI targets, an investment in the entire grid is not needed. LV monitoring is helpful to detect fraud, with a higher granularity and detection easiness if meter to feeder connectivity is deployed.

The following criteria will be used to select the telecommunication solution for each substation category and type:

- Wireline technologies are preferred to wireless technologies, as the former provide a higher predictability and availability.

- Broadband solutions are preferred to narrowband solutions, as broadband can provide the bandwidth needed to create a multiservice network.

- Private network technologies are preferred to TSP services, as they provide a close control of the service level provided. However, investment and recurring costs [i.e., total cost of ownership (TCO)] analysis need to be made considering the expected lifespan of the technology or the associated terminals.

The TCO of the telecommunication solution in an SS (the access network part, including its core network share), is an important informational driver for technology selection. This cost should be compared with the cost of the IEDs and application servers and network share in each substation. In a perfect world where all technologies could be available for any substation, the cost percentage of telecommunications referred to the total investment in the substation for the rest of the smart grid related elements is an input of the decision. For example, a higher IED investment could allow a higher cost telecommunication technology selection in specific SSs; by contrast, if the lowest-cost available technology is not available in the substation (e.g., radio coverage), a higher-cost, but readily available, option will be selected:

- BASS category will not easily justify high-cost telecommunication technologies. This type of substation can be one in a rural area where MV automation would not be useful due to the existing MV infrastructure, or one with a low count of meters connected to the substation.

- LMSS and MMSS categories are associated to real-time telecommunication technologies and moderate bandwidth requirement, where latency is not a major constraint.

- AUSS category involves real-time, good bandwidth, low latency and controlled jitter telecommunication services. This substation type is always related to mission-critical needs, and increased availability and reliability. The cost of the smart grid-related assets is high and usually justifies a high investment on a multiservice telecommunication network that can differentiate and prioritize different data flows.

- The ASSS attribute is associated with continuous high-data transfer need. A multiservice broadband telecommunication network, where this type of traffic can be made of less priority than others, is advisable.

9.3.2   High-Level Architecture

Figure 9.3 shows the different systems and how they connect among them, with the IEDs in the SSs and the smart meters through the telecommunication network.

From an application perspective, the AMI system includes the connection to the point-of-supply information system so that the consumption information may be provided to retailers, and that power limitation, and the connection/disconnection orders may be executed over the smart meters. The DA system is an evolution of a traditional SCADA. It is supported over the same telecommunication network and integrates the operation of the LV grid through a connection with the AMI system. It does not extend to all SSs but to about 20% of them.

The telecommunication network to support AMI and DA is organized as proposed by Chapter 6 with the core, access, and local network blocks. The core network is organized around a Multiprotocol Label Switching (MPLS) network with a limited number [a few tens of nodes implementing virtual private routing network (VPRN) services to isolate different data flows]. The access network combines a number of private and public telecommunication technologies including BPL, Ethernet transport over optical fiber cables, 2G/3G public

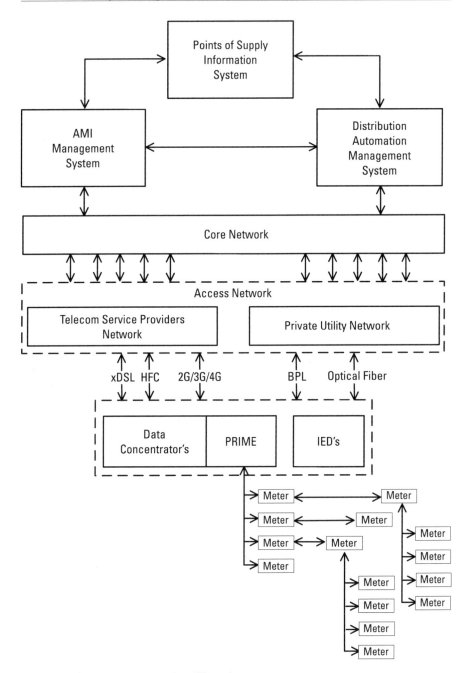

Figure 9.3 System interconnection, IED, and smart meter access.

radio solutions (GPRS and UMTS), and ADSL or HFC connectivity. The local network is based on ITU-T Recommendation G.9904 (PRIME) PLC connectivity in CENELEC-A band as specified by ITU-T Recommendation G.9901, integrated in the smart meters and the AMI DC.

The access network is the most challenging network block, providing access to tens of thousands of SSs. This access network combines private and public nature telecommunications, in an attempt to balance end-to-end performance, the TCO, and the future evolution capabilities. From a technology perspective, private BPL solutions, together with segments of optical fiber cables, are combined with TSP-provided 2G/3G radio solutions, and ADSL or HFC fixed network access, as in Figure 9.4. Table 9.2 shows the different solution percentages of a total of more than 35,000 deployed SSs.

Some of the SSs are accessed individually with Ethernet over optical fiber, 2G/3G public radio, ADSL, or HFC technologies. Ethernet over optical fiber is part of the private utility network with a very high performance, both in terms of throughput and latency; it is implemented with regular Ethernet switches

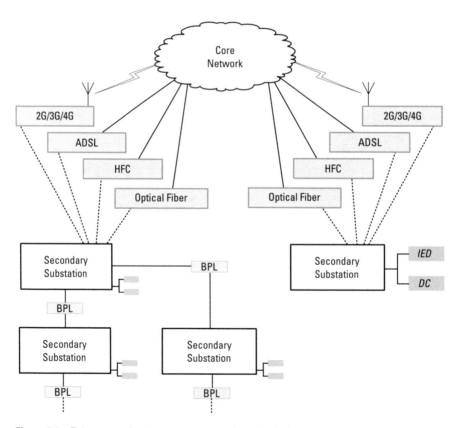

Figure 9.4 Telecommunication access network technologies.

Table 9.2
Access Network Telecommunication Technologies Deployed in the Use Case

Solution Category	Total Number of SS (%)	Subtotal Number of Each SS Solution Category (%)
Individual SS	60	
Public 2G/3G radio		97
Public ADSL/HFC		1
Private Ethernet over optical fiber		2
Cluster of SSs (BPL)	40	
Backbone: Public 2G/3G radio		33
Backbone: Public ADSL/HFC		37
Backbone: Private Ethernet over optical fiber		30

connected via optical fiber configuring rings. 2G/3G radio network is deployed with a single radio but double SIM card to connect either to two different TSPs; the tariff model is based on a per-volume payment. ADSL and HFC are wireline connections with a flat-rate tariff model.

BPL is used to provide L2 connectivity among a group of SSs ("cluster"; administratively limited to a maximum of 20 substations) using OPERA technology (see Chapter 5). Each BPL cluster has a backbone transport connectivity to link the cluster with the central application servers. This transport connectivity is provided either with Ethernet over optical fiber, 2G/3G public radio, ADSL, or HFC (in some cases, where redundancy is needed, a secondary backbone connection is included [3]). The clusters are designed to achieve a minimum of 100 kbps bidirectional application layer throughput with all SSs in a BPL cluster transmitting and receiving traffic simultaneously.

9.3.3 Telecommunication Network Dimensioning

The main telecommunication design targets in this use case are summarized in Table 9.3. They are realistic, cost-efficient, and future-proof.

9.3.4 Deployment Process

A smart grid deployment process is a set of activities defined and executed to define, purchase (order and receive), install, check, and accept the smart grid solutions in the SSs, prior to their normal operation through the systems. The deployment of telecommunication-related devices is considered simultaneous to the deployment of the IEDs and other auxiliary elements related to the smart

Table 9.3
Requirements of the Different Telecommunication Segments

	Local Network (PRIME PLC)	Access Network (shared connection with the secondary substations)		Access Network (ACR connection to Core Network)	Core Network (Core Router connectivity)
	AMI	AMI	DA		
Data rate	20 kbps	20–100 kbps		2–10 Mbps	1–10 Gbps
Latency	2–15 sec	2 sec	100 ms–2 sec	100 ms	10 ms
Packet loss	5%	1–2%		0.1%	0.01%
Reliability	Low	Medium	Medium-high	High	Very high
QoS	Not needed	If possible		Mandatory	Mandatory
Backup Power	Not needed	Not needed	6–24 hours	24 hours	72 hours
Security	High	High		High	Very high
Scalability	Low	Medium	Medium	High	Very high

grid and not in the telecommunications' domain. It will be assumed that they will be installed in the same rack, cabinet, or enclosure.

The deployment process is summarized in Table 9.4. The focus of the deployment in terms of telecommunications is placed on the SS access network. The field survey, the telecommunication network design and the preproduction test will be further detailed.

9.3.4.1 Secondary Substation Field Surveys and Telecommunication Network Planning

Telecommunication network planning is one step of the overall process, but a very important one to assign the best possible solution in each SS. It must be completed within the given constraints imposed from the economic and the technical limits associated to the different telecommunication solutions available within the utility strategy. The "initial telecommunication network planning" refers to the definition of the telecommunication solution in each SS with the information available in the company information systems.

If the telecommunication solution is based on public services, the information from the different TSP's service availability must be checked at each site (SS). For example, if ADSL is to be used, the footprint of the commercial service must be checked with the TSP; if the telecommunications are going to be based on 2G/3G radio, theoretical coverage is needed (in the case of underground or urban sites, a certain extra power loss should be considered). If the telecommunication solution is going to be based in private telecommunication systems, the utility should revert to its own planning tools and methodology.

Field surveys are important to analyze the constraints of the SSs. Apart from the electricity-related aspects that need to be checked on field, the survey

Table 9.4
Smart Grid Deployment Steps

Stage	Activity
Functionality definition and field surveys	List of SSs to be deployed in an area
	Definition of the category of each SS (i.e., functional definition)
	Initial telecommunication network planning
	Definition of the deployment phases in each deployment area
	Field survey for each SS in the selected deployment phase
	Final telecommunication network planning, after the analysis of the survey data
	Metering rooms field survey (safety conditions)
	Deployment phase scheduling definition
Systems arrangements	Data validation
	Selection of equipment, enclosure assembler and field installer for each SS
	IP addressing assignment
	Purchase orders to selected enclosure assemblers and field installers
Equipment manufacturing	Manufacturing of equipment and enclosure assembly
	Equipment configuration
	Factory tests, including individual devices and assembled enclosure
	Delivery of enclosures to field installers
Equipment on field installation	Acceptance of cabinets by field installers
	Field SS installation
	Installation tests
Preproduction tests	Telecommunication preproduction test of the access network
	Smart meter deployment completion in the deployment area
	Telecommunication preproduction test of the local network (smart meters connectivity)
	Individual application preproduction tests
	Qualification of the deployment area

should consider some key aspects that will be used to create a "final telecommunication network planning." The aspects that should be checked are:

- Availability of any preexisting telecommunication solution, private network, or public service, that could be already installed or have previously been present.
- Availability of xDSL, HFC, or FTTx infrastructure in the vicinity of the SS (e.g., cable distribution boxes).
- Coverage of 2G/3G/4G radio networks of the different TSPs in the service territory. The coverage should consider the parameters needed to

check it (received power from different base stations, RSSI, in GPRS, but other values such as E_c/N_0 and RSCP in UMTS), and should be measured inside (with all doors closed) and outside the substation with a reference antenna.

- Cable entry points in the SS; distances for the different cables needed in the substation (radio connection antenna, cables for the services, cables for the PLC connectivity, and so forth).
- Best location for the antenna, inside or outside the substation.
- MV switchgear types to decide on the best BPL couplers.
- LV busbar connection points for PLC.

Once all the field information is available, the final telecommunication network planning can be prepared. Any constraint related to the technology should be considered at this stage to schedule the deployment of the individual SSs.

9.3.4.2 Telecommunication Preproduction Tests

Preproduction tests refer to the tests performed prior to the production phase (i.e., before any deployed system is delivered to be operated in normal conditions). This terminology is derived from software development environments, and it is quite convenient to consider these tests for a complex system as the smart grid.

Preproduction telecommunication network tests are performed to verify that the performance results are not only enabling the smart grid services, but are according to the performance expectations as well. The results serve as the benchmark against any future diversion from the registered results:

- Test period should be long enough to capture the long-term performance of the solution. A period of 1 week could be considered optimal; however, as a trade-off, a period of 2 days (48 hours) will be considered adequate for the intended objective.
- Multiple performance parameters could be measured. However, given a known traffic pattern (in this case under analysis, the traffic pattern will be the one of DA, with short-length packets, below 256 bytes, and predictable latency) these tests can consider the results in terms of packet loss and round-trip delay time (RTT as in IPPM; framework defined in RFC 2330).
- Depending on each telecommunication technology, there is a different expected performance and, thus, the acceptance thresholds should be different. These are shown in Table 9.5.

Table 9.5
Performance Thresholds

Technology Combination	% Maximum Packet Loss	Average Round-Trip Delay Time (ms)
Individual SS with public 2G	1.40	1,500
Individual SS with public 3G	1.40	1,000
Individual SS with public ADSL/HFC	0.75	150
Individual SS with private Ethernet over optical fiber	0.40	30
Cluster of BPL-connected SSs with public 2G backbone. Targets for the 5 closest SSs to the ACR	1.60	1,575
Cluster of BPL-connected SSs with public 2G backbone. Targets for the SSs further from the ACR than 5 SSs	1.80	1,650
Cluster of BPL-connected SSs with public 3G backbone. Targets for the 5 closest SSs to the ACR	1.60	1,075
Cluster of BPL-connected SSs with public 3G backbone. Targets for the SSs further from the ACR than 5 SSs	1.80	1,150
Cluster of BPL-connected SSs with public ADSL/HFC backbone. Targets for the 5 closest SSs to the ACR	0.95	225
Cluster of BPL-connected SSs with public ADSL/HFC backbone. Targets for the SSs further from the ACR than 5 SSs	1.15	300
Cluster of BPL-connected SSs with private Ethernet over optical fiber backbone. Targets for the 5 closest SSs to the ACR	0.60	105
Cluster of BPL-connected SSs with private Ethernet over optical fiber backbone. Targets for the SSs further from the ACR than 5 SSs	0.80	180

Table 9.6 shows the results of the preproduction tests of some examples of connectivity in different SSs (Figure 9.5). The test pattern consists of a sequence of ICMP pings with 256 byte packets sent every minute.

9.3.5 Telecommunication Network Management

Telecommunication network management is the way to guarantee that the network is operational and performs as expected. At the same time, the telecom-

Table 9.6
Performance Results of the Example Substations

Technology	% Packet Loss	RTT (ms)
Individual SS with public 3G	0.73	500.15
Individual SS with private Ethernet over optical fiber	0.00	7.51
Cluster of BPL-connected SSs with private Ethernet over optical fiber backbone	0.52	152.66

RTT (ms) 3G Public Radio Connect Secondary Substations

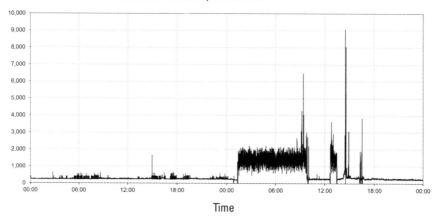

RTT (ms) Private BPL Connect Secondary Substations

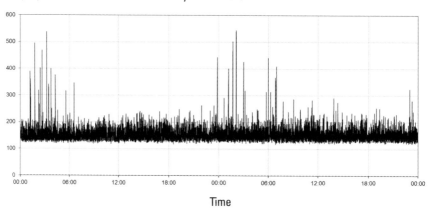

RTT (ms) Private Ethernet Optical Connected Secondary Substations

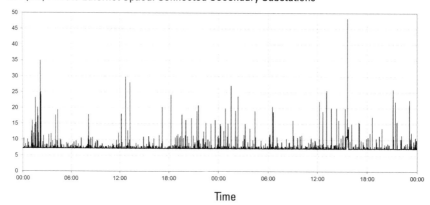

Figure 9.5 RTT (30 seconds average of ICMP pings every 5 seconds) of several example secondary substations for 2 days.

munication network management is the tool to coordinate activities on the field.

Telecommunication network management cannot be performed if the proper tools are not in place, and there is not a group of engineers and technicians that through them operate and maintain the network; this group of resources is typically referred to as network management center (NMC) or network operation center (NOC). The network is managed associating graphical interfaces representing geography and assets to direct field operations, and the different network blocks are managed as independent layers and connected through the proper interfaces; different network layers with varied complexity levels are operated with separated tools and a group of technicians.

9.3.5.1 Telecommunication Network Metrics and Application Performance Indicators

The grid reliability is monitored with a set of indices (SAIFI, SAIDI, and so forth). They basically measure the distribution grid availability. The advent of the smart grid intends to improve this performance metrics, but also others (e.g., quality related), with the new applications it brings along. The collection of real-time grid information, its analysis, the remote control of grid elements, and the automatic grid recovery procedures will be instrumental for this purpose.

It is not clear which will be the impact of each of the smart grid applications in the improvement of the power grid service itself, but it is obvious that these new applications need to perform correctly. The telecommunication network performance is the foundation to ensure that the applications can work to the best of their possibilities. Although a mathematical correlation between any application and the telecommunication supporting it cannot be drawn in a general way, a direct correlation exists (see Figure 9.6).

Telecommunication network performance can be monitored independently of the applications it supports. However, a better derivation of performance indicators can be made if the application performance indicators are known (see Table 9.7).

Application-level performance, in case it is below expectations, is just the symptom that something is not working properly. The root cause must be derived, and telecommunication network-level indicators must be observed to find if the origin lies on the telecommunication network. Insufficient available bandwidth, excessive latency, or excessive packet loss are common metrics that could help to identify telecommunication network-related problems. For example, if the problem with a certain IED is in the abnormal latency, this could come from its software implementation, or a throughput bottleneck either at the IED telecommunication connection or at a certain network interface; it could also have its origin in a packet asymmetric routing or synchronization

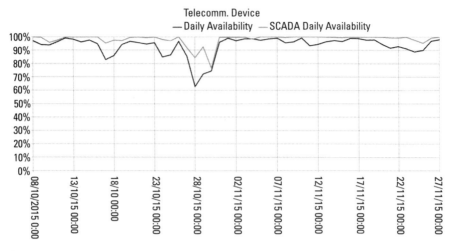

Figure 9.6   Telecommunication and application-level availability for a 3G router.

Table 9.7
Typical Performance Indicators of AMI and DA Systems

AMI	DA
Total number of meter readings and load curves retrieved	Availability of IEDs
Last retrieval time of the meter readings and load curves	Read success rate of IED's parameters
Remote operations success rate: Power limitations, instant readings, connection and disconnection of electricity service, contract modifications	Remote operations success rate: Open/close actions (e.g., switchgears), instant parameters readings
LV grid and consumption related events received	
Availability of data concentrators	

problems. Network probes for traffic analysis, as part of the telecommunication network management tools may help to narrow the problem (Figure 9.7).

From a smart grid perspective, application and telecommunication performance indices may also be blended [4]:

- Bandwidth: Data rates, at both telecommunication network and application levels;
- Timeliness: Latency, jitter and response time;
- Availability: Uptime and recovery time (from unavailable time);
- Loss: Lack of information or erroneous information reception;

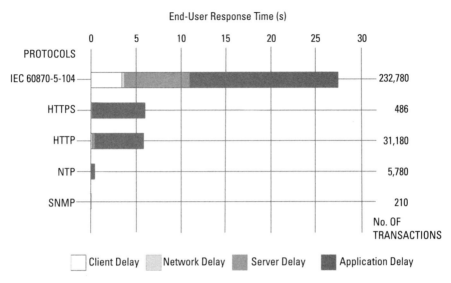

Figure 9.7   Latency study in a secondary substation connection.

- Security and trust: Confidentiality, integrity, nonrepudiation, and authentication.

These metrics need to be followed in a systematic way. The following specific management activities can be referenced:

- Monitoring of equipment availability: Equipment refers here to network nodes and to end devices. This monitoring is based on the ability to detect and locate connectivity alarms in the devices, either collecting traps (SNMP) sent by them or via periodic polling processes. In the latter case, depending on the criticality of the services (end devices) or the importance of the network node, different polling frequencies need to be defined (1 minute, 5 minutes, 1 hour, and so forth).
- Monitoring of equipment-critical internal components: This activity involves monitoring CPU consumption, memory use, and temperature to check device health.
- Monitoring of sets of basic parameters of the core, access, and local network devices (e.g., packet loss and latency): The activity here involves checking diversion from historical trends.
- Monitoring of sets of basic parameters associated to specific technologies used in different network blocks: For example, if radio technologies

are used, the coverage level and its evolution are helpful; if optical links are used, received optical power needs to be monitored.

- Traffic monitoring per SS: This activity controls normal traffic consumption patterns (especially important when TSP's services are used).
- Traffic monitoring at relevant network interfaces and links: The link utilization at the relevant network points (e.g., aggregation points, interconnections between different networks, critical paths in the transport network) must be studied to control network growth.
- DPI of traffic at relevant network interfaces, controlling the network use by the applications to identify potential problems and improve the network design and dimensioning.

9.3.5.2 Smart Grid Network Traffic

There is a strong correlation between the architecture design and the performance of end-to-end applications. The telecommunication network is designed to cope with the application requirements, and applications must respect those initial requirements. However, initial (sometimes theoretical) requirements are not always well defined and eventually change over time; the telecommunication network should be able to evolve accordingly to accommodate traffic growth and pattern changes.

This section shows the throughputs registered at various relevant points of the network as a way to quantify traffic demand. These data rates are the most visible metric in this use case, as AMI is very tolerant to latency and jitter parameters, and DA can be adjusted to the worst case expected latency, absorbing variations in the end-to-end delay. The following aspects must be considered:

- Apart from the AMI and DA traffic, there is extra traffic for the control and operation of the telecommunication network (mainly protocol overheads and the network availability checking processes).
- Two types of SSs will be shown:
  - *Individually telecommunicated substations:* A single router handles just the traffic from this substation.
  - *Substations telecommunicated as part of a cluster of interconnected substations:* At one of the substations, a router aggregates the traffic of a set of SSs within the same L2 domain.
- Graphs are derived from a SNMP-based network management tool retrieving transmission and reception WAN interface counters from network elements every 5 minutes. Throughputs are consequently aver-

ages over that period, and instant traffic peaks are not reflected. TCP/IP mechanisms manage to adjust traffics to the available links' throughput.

Individually Connected Secondary Substation in the Access Network

Figure 9.8 shows the throughput of an SS with an IEC 60870-5-104 IED and an AMI DC with 62 PRIME PLC smart meters. It can be observed that throughput is higher upstream than downstream due to the AMI; the traffic pattern, in the absence of abnormal network conditions, has a periodic nature.

Figure 9.9 shows the traffic of the DA and AMI services connected. The difference of the peak values between Figure 9.9 compared to Figure 9.8 is the averaging of data for storage purposes of the tool used to monitor traffic.

AMI DC traffic shows peaks starting at midnight when smart meter readings and load profiles are scheduled to be sent to the application servers via FTP (upload traffic is higher than the download traffic). When other peaks appear, they are due to specific schedules (e.g., smart meters event transfer) or non-scheduled operations (firmware upgrades, instant meter readings, and so forth). DA traffic in the IED shows a normally stable behavior, close to symmetric and with scarce and low peaks. AMI traffic is the most significant contribution to the total traffic flow.

The asymmetric nature of the traffic is an important consideration for network dimensioning. In this case, and common to smart grid networks, upstream traffic is higher than the downstream traffic (as opposed to traditional Internet use). Under normal operation conditions, upstream needs to

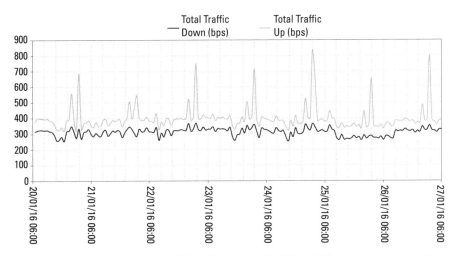

Figure 9.8   Data throughput of a 2G/3G radio router with AMI and DA services in a secondary substation in 1 week.

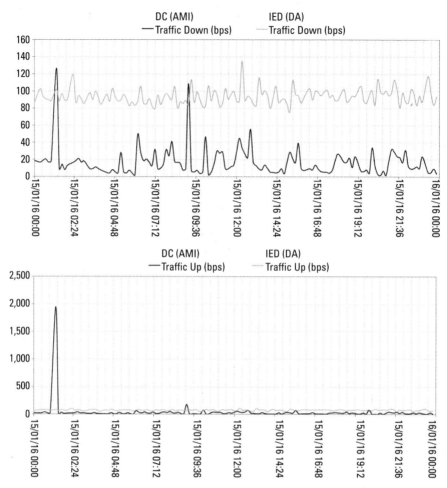

Figure 9.9 Data throughput of the DA IED and AMI DC services in a secondary substation.

be considered as the most constraining traffic in the definition of the network bandwidth. However, there are operation conditions that do not follow this pattern, such as IED or network equipment firmware download. Thus, not only normal operation traffic conditions must be used to design the needed network bandwidth, but also other nondaily operations that may offer another network traffic pattern. Otherwise, for example, massive unicast firmware downloads may not be possible simultaneously, and the smart grid applications or servers must control that the network is used without causing congestion. Moreover, bandwidth-constrained access network links may be fully booked by this massive traffic demand, and if the telecommunication technology used does not support QoS, critical applications may become unavailable during the time the congestion conditions exist.

It is also important to mention that, apart from the smart grid operations controlled by automatic application tasks, human intervention is also expected. Automatic tasks are the ones that offer the predictable periodic traffic patterns. Examples of manual operations can be instant meter readings to particular meters, meter connect and disconnect orders, and IED parameter retrieval for verification purposes.

Cluster of Interconnected Secondary Substations in the Access Network

The traffic volume in the aggregation point of a cluster of substations adds the individual traffics, and includes the extra-traffic mentioned above (e.g., the aggregation is using DMVPN and IPSec in transport mode). Figure 9.10 shows the data throughput of a BPL cluster of 5 SSs, with 5 AMI DCs and a total of 654 smart meters and 2 DA IEDs. The different peaks starting at midnight show the effect of the 5 AMI DCs uploading data at different times, due to the different time it takes for each to poll the diverse number of smart meters in each SS.

Core Network Links

All network traffic is finally routed by the core network to its destination. Thus, the core network needs to absorb all the traffic created by the smart grid services at the end-points. The different application servers' networks are connected to the core network, and the different TSPs' interconnection interfaces are attached. These interfaces must also be dimensioned properly not to create bottlenecks.

Figure 9.11 shows the traffic flowing in one representative connection between a pair of core routers for 1 week; this link carries also nonsmart grid

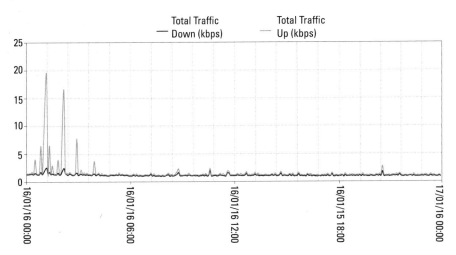

Figure 9.10 Aggregated data throughput of a BPL cluster of 5 secondary substations.

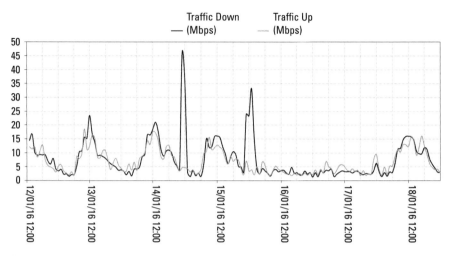

Figure 9.11   Traffic between core routers.

traffic associated to corporate offices, and that is the explanation for the decrease in the traffic on the weekend.

Figure 9.12 shows the service traffic delivery in a day at one of the interconnection points of the AMI and DA application servers (representing a 18% of the total SSs; AMI and DA traffic is delivered in an aggregated way). Figure 9.13 shows the same traffic over 1 year, in which the number of SSs deployed have been doubled.

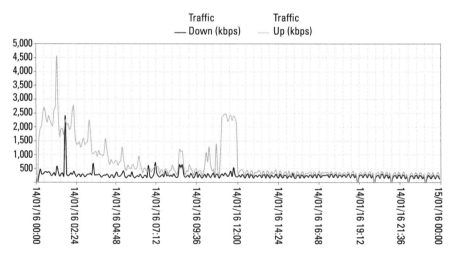

Figure 9.12   Daily traffic in one of the AMI and DA network interfaces for smart grid service delivery.

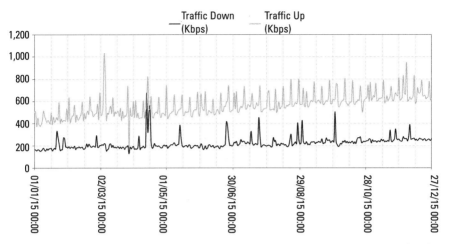

Figure 9.13 Traffic in one of the AMI and DA network interfaces for smart grid service delivery in 1 year.

Public Network (TSP) Connection Interfaces

As mentioned in the previous section, if TSPs are used there are interfaces through which the two networks, the private and the public ones, get connected. These interfaces are the traffic exchange points, and are typically implemented with redundancy and sometimes load balancing mechanisms. Two physical interfaces in different distant locations and load balancing are present in this use case.

Figures 9.14 and 9.15 (the peak comparison is not equal due to the averaging of the tool used, for data storage purposes) show the evolution of the

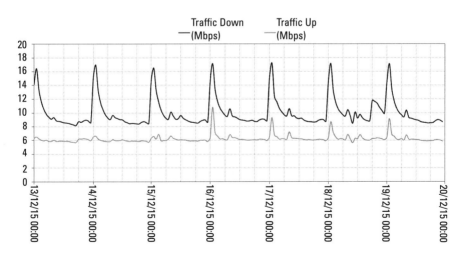

Figure 9.14 Total traffic in the public network (TSP) connection interface in 1 week.

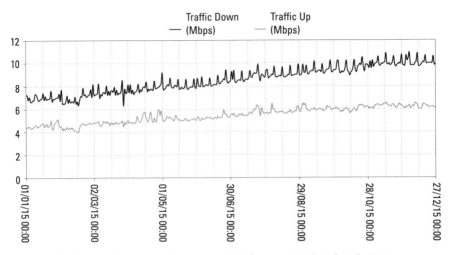

Figure 9.15 Total traffic in the public network (TSP) connection interface in 1 year.

traffic in the logical interface (all physical interfaces aggregated) of this use case, in the periods of 1 week and 1 year, respectively. The growth in 1 year has been of 80% more SSs. The AMI meter values upload starting at midnight drives the traffic pattern.

The study of the traffic evolution is crucial to know the real demanded traffic and make the network links and equipment evolve to absorb the traffic flow, often different from initial estimations.

References

[1]   Energy Technology Policy Division, *Technology Roadmap: Smart Grids,* Paris: International Energy Agency (IEA), 2011. Accessed March 27, 2016. http://www.iea.org/publications/freepublications/publication/smartgrids_roadmap.pdf.

[2]   Sendin, A., et al., "PLC Deployment and Architecture for Smart Grid Applications in Iberdrola," *Proc. 2014 18th IEEE International Symposium on Power Line Communications and Its Applications,* Glasgow, March 30–April 2, 2014, pp. 173–178.

[3]   Solaz, M., et al., "High Availability Solution for Medium Voltage BPL Communication Networks," *Proc. 2014 18th IEEE International Symposium on Power Line Communications and Its Applications,* Glasgow, March 30–April 2, 2014, pp. 162–167.

[4]   Goel, S., S. F. Bush, and D. Bakken, (eds.), *IEEE Vision for Smart Grid Communications: 2030 and Beyond,* New York: IEEE, 2013.

List of Acronyms

1G First Generation Mobile Network

2G Second Generation Mobile Network

3DES Triple Data Encryption Standard

3G Third Generation Mobile Network

3GPP Third Generation Partnership Project

3GPP2 Third Generation Partnership Project 2

4G Fourth Generation Mobile Network

4W E&M 4 Wires Ear and Mouth

6LoWPAN IPv6 over low-power wireless personal area networks

8PSK Eight-Phase Shift Keying

AAAC All-Aluminum Alloy Conductor

AAC All-Aluminum Conductor

AACSR Aluminum alloy conductor steel reinforced

ac Alternating current

ACAR   Aluminum conductor alloy reinforced

ACK   Acknowledgment

ACR   Access connection router

ACSR   Aluminum conductor steel reinforced

ACSS   Aluminum conductors steel supported

ADA   Advanced distribution automation

ADM   Add-drop multiplexer

ADPSK   Amplitude differential phase shift keying

ADSL   Asymmetric digital subscriber line

ADSS   All-dielectric self-supporting

AES   Advanced Encryption Standard

AM   Amplitude modulation

AMI   Advanced metering infrastructure

AMPS   Advanced mobile phone service

AMR   Automatic meter reading

ANSI   American National Standards Institute

AON   Active optical network

APCO   Association of Public-Safety Officials

API   Application programming interface

APN   Access point name

ARIB   Association of Radio Industries and Business (Japan)

ARP   Address Resolution Protocol

ARQ   Automatic repeat request

ATIS   Alliance for Telecommunications Industry Solutions

ATM   Asynchronous transfer mode

AUC   Authentication center

AWG   American Wire Gauge

BAN   Building area network

BB   Broadband

BEMS   Building energy management system

BGP   Border Gateway Protocol

BN   Base node

BPL   Broadband power line

BPON   Broadband passive optical network

bps   Bits per second

BPSK   Binary phase shift keying

BSS   Basic service set

BTS   Base transceiver station

BWA   Broadband wireless access

CAD   Consumer access device

CAIDI   Customer Average Interruption Duration Index

CAIFI   Customer Average Interruption Frequency Index

CAN Controller area network

CATV Community antenna television

CCTV Closed-circuit television

CDMA Code division multiple access

CEN Comité Européen de Normalisation (European Committee for Standardization)

CENELEC European Committee for Electrotechnical Standardization (Comité Européen de Normalisation Electrotechnique)

CEPT European Conference of Postal and Telecommunications

CFP Contention-free period

CIGRE Conseil International des Grands Reseaux Electriques (International Council on Large Electric Systems)

CIR Committed information rate

CISPR Comité International Spécial des Perturbations Radioélectriques (International Special Committee on Radio Interference)

CLI Command line interface

CM Cable modem

CMTS Cable modem termination system

CoAP Constrained Application Protocol

COSEM Companion Specification for Energy Metering

CPE Customer premises equipment

CPP Critical peak pricing

CPR Critical peak rebates

CRC   Cyclic redundancy check

CRM   Customer relationship management

CSMA   Carrier sense multiple access

CSMA/CA   Carrier sense multiple access with collision avoidance

CSMA/CD   Carrier sense multiple access with collision detection

CSP   Communications service provider

CTCSS   Continuous Tone-Controlled Squelch System

CTS   Clear to send

CWDM   Coarse wavelength division multiplexing

D8PSK   Differential eight-phase shift keying

DA   Distribution automation

DBPSK   Differential binary phase shift keying

dc   Direct current

DC   Data concentrator

DCC   Data communications company

DECC   Department of Energy & Climate Change

DER   Distributed energy resources

DES   Digital Encryption Standard

DG   Distributed generation

DHCP   Dynamic Host Configuration Protocol

DL   Downlink

DLC   Distribution line carrier

DLL   Data link layer

DLMS   Device Language Message Specification

DLR   Dynamic line rating

DMR   Digital mobile radio

DMS   Distribution management system

DMT   Discrete multitone

DMVPN   Dynamic multipoint virtual private network

DNP3   Distributed Network Protocol

DOCSIS   Data Over Cable Service Interface Specification

DOE   Department of Energy (U.S.)

DP   Dynamic pricing

DPD   Dead peer detection

DPI   Deep packet inspection

DQPSK   Differential quadrature phase shift keying

DR   Demand response

DS   Distributed storage

DSL   Digital subscriber line

DSLAM   Digital subscriber line access multiplexer

DSM   Demand side management

DSO   Distribution system operator

DSP   Data service provider

DSR   Demand side response

DSSS   Direct sequence spread spectrum

DTLS   Datagram Transport Layer Security

DWDM   Dense wavelength division multiplexing

EC   European Commission

EDGE   Enhanced Data rates for Global Evolution

EGP   Exterior Gateway Protocol

EHF   Extremely high frequency

EIA   U.S. Energy Information Agency

EIGRP   Enhanced Interior Gateway Routing Protocol

EIR   Excess information rate

EIRP   Equivalent isotropically radiated power

EISA   Energy Independence and Security Act (U.S.)

EMC   Electromagnetic compatibility

EMF   Electromagnetic field

EMS   Energy management system

EN   European Standard

ENISA   European Union Agency for Network and Information Security

EPON   Ethernet passive optical network

EPR   Ethylene propylene rubber

EPRI   Electric Power Research Institute

ERP   Equivalent radiated power, Enterprise resource planning

ESD   Electrostatic discharge

ESO   European Standard Organization

ETSI   European Telecommunications Standards Institute

EU   European Union

EV   Electric vehicle

FACTS   Flexible Alternating Current Transmission System

FAN   Field area network

FCC   Federal Communications Commission (U.S.)

FDMA   Frequency division multiple access

FEC   Forward error correction

FHSS   Frequency-hopping spread spectrum

FM   Frequency modulation

FR   Frame relay

FRR   Fast reroute

FSK   Frequency shift keying

FTP   File Transfer Protocol

FTTx   Fiber To The x

GAN   Global area network

GEO   Geostationary (or geosynchronous) Earth orbit

GIS   Geographic information system

GOOSE   Generic object-oriented substation event

GOS   Grade of service

GPON   Gigabit passive optical network

GPRS   General packet radio service

GPS   Global positioning system

GRE   Generic routing encapsulation

GSM   Global System for Mobile

GUI   Graphical user interface

GW   Gateway

HAN   Home area network

HDR   High data rate

HDSL   High bit rate digital subscriber line

HE   Head-end

HEMS   Home energy management system

HES   Home electronic system

HF   High frequency

HFC   Hybrid fiber-coaxial

HLR   Home location register

HSDPA   High-speed downlink packet access

HSPA   High-speed packet access

HSUPA   High-speed uplink packet access

HV   High voltage

HVAC   Heating, ventilation, and air conditioning

ICMP   Internet Control Message Protocol

ICT   Information and communication technologies

IDPS   Intrusion detection and prevention system

IDS   Intrusion detection system

IEC   International Electrotechnical Commission

IED   Intelligent electronic device

IEEE   Institute of Electrical and Electronics Engineers

IETF   Internet Engineering Task Force

IMT   International Mobile Telecommunications

IoT   Internet of Things

IP   Internet Protocol

IPPM   Internet Protocol Performance Metrics

IPS   Intrusion prevention system

IPsec   Internet Protocol security

IPv4   Internet Protocol version 4

IPv6   Internet Protocol version 6

ISDN   Integrated services digital network

IS-IS   Intermediate system to intermediate system

ISM Industrial, scientific, and medical

ISO International Organization for Standardization

IT Information technology

ITU International Telecommunication Union

ITU-R International Telecommunication Union – Radiocommunication Sector

ITU-T International Telecommunication Union – Telecommunication Standardization Sector

L2 Layer 2

L3 Layer 3

LAN Local area network

LCP Low complexity profile

LDAP Lightweight Directory Access Protocol

LDP Label Distribution Protocol

LDPC Low-density parity check

LED Light-emitting diode

LEO Low Earth orbit

LF Low frequency

LLC Logical link control

LMR Land mobile radio

LOS Line of sight

LSP Label switched path

LTE   Long-term evolution

LV   Low voltage

M2M   Machine-to-machine

MAC   Medium access control

MAIFI   Momentary Average Interruption Frequency Index

MAN   Metropolitan area network

MCM   Multicarrier modulation

MCS   Modulation and coding schemes

MDM   Meter data management

MDMS   Meter data management system

MDU   Multiple dwelling unit

MEMS   Microgrid energy management system

MEO   Medium Earth orbit

MF   Medium frequency

MIB   Management information base

MICE   Mechanical, ingress, climatic/chemical, electromagnetic

MOS   Mean Opinion Score

MP   Multiprotocol

MPLS   Multiprotocol Label Switching

MPLS-TP   Multiprotocol Label Switching-Transport Profile

MR   Multiregional

MTBF   Mean time between failure

MTC   Machine-type communications

MTPAS   Mobile Telecommunications Privileged Access Scheme

MTTF   Mean time to failure

MTTR   Mean time to repair

MV   Medium voltage

MVNO   Mobile virtual network operator

NAN   Neighborhood area network

NAT   Network address translation

NB   Narrowband

NEMA   National Electrical Manufacturers Association

NERC   North American Electric Reliability Corporation (USA)

NETL   National Energy Technology Laboratory (USA)

NHRP   Next Hop Resolution Protocol

NIST   National Institute of Standards and Technology (USA)

NLOS   Nonline of sight

NMC   Network management center

NMS   Network management system

NOC   Network operation center

NSO   National Standards Organization

NTP   Network Time Protocol

O&M   Operation and maintenance

OADM   Optical add-drop multiplexers

OAM   Operations, administration, and maintenance

OCh   Optical channel

ODN   Optical distribution network

ODU   Optical data unit

OFCOM   Office of Communications (UK)

OFDM   Orthogonal frequency division multiplexing

OFDMA   Orthogonal frequency division multiplexing access

OLT   Optical line terminations

OMS   Outage management system

ONT   Optical network termination

OPERA   Open PLC European Research Alliance

OPGW   Optical ground wire

OPPC   Optical phase power cable

OQPSK   Offset quadrature phase shift keying

OSC   Optical supervisory channel

OSI   Open system interconnection

OSPF   Open shortest path first

OTN   Optical transport network

OTU   Optical transport unit

OXC   Optical cross-connect

PAMR   Public access mobile radio

PAN   Personal area network

PC   Personal computer

PDC   Phasor data concentrator

PDH   Plesiochronous digital hierarchy

PDU   Packet data unit

PE   Polyethylene

PHY   Physical

PILC   Paper insulated lead covered

PKI   Public key infrastructure

PLC   Power line communication (or carrier)

PLT   Power line telecommunication

PMR   Private (or professional) mobile radio

PMU   Phasor measurement unit

PON   Passive optical network

POTS   Plain Old Telephone Service

PPDR   Public Protection and Disaster Relief

PRIME   Powerline for Intelligent Metering Evolution

PRM   PHY robustness management

PS   Primary substation

PSD   Power spectral density

PSK   Phase shift keying

PSTN   Public switched telephone network

PTT   Push-to-talk

PV   Photovoltaic

PVC   Polyvinyl chloride

QAM   Quadrature amplitude modulation

QC-LDPC-BC   Quasi-cyclic LDPC block code

QoS   Quality of service

QPSK   Quadrature phase shift keying

RADIUS   Remote Authentication Dial In User Service

RF   Radio frequency

RFC   Request for Comment

RFID   Radio frequency identification

RG   Radio guide

RIP   Routing Information Protocol

rms   Root mean square

ROADM   Reconfigurable optical add-drop multiplexer

RPL   Routing Protocol for low power and lossy networks

RR   Radio Regulations

RSCP   Received signal code power

RSSI   Received signal strength indication

RSTP   Rapid Spanning Tree Protocol

RSVP   Resource Reservation Protocol

RSVP-TE   Resource Reservation Protocol-Traffic Engineering

RTP   Real-time pricing

RTS   Request to Send

RTT   Round-trip delay time

RTU   Remote terminal unit

SA   Substation automation

SAIDI   System Average Interruption Duration Index

SAIFI   System Average Interruption Frequency Index

SAN   Storage area network

SAR   Segmentation and reassembly

SAS   Substation automation system

SCADA   Supervisory control and data acquisition

SCTE   Society of Cable Telecommunications Engineers

SDH   Synchronous digital hierarchy

SDO   Standards development organization

SEC   Smart energy code

SEP   Smart energy profile

SGIP   Smart Grid Interoperability Panel

SHDSL   Single-pair high-speed digital subscriber line

SHF   Super high frequency

SIM   Subscriber identity module

SLA   Service level agreement

SMETS2   Smart Metering Equipment Technical Specifications: second version

SMHAN   Smart metering home area network

SMWAN   Smart metering wide area network

SN   Service node

SNMP   Simple Network Management Protocol

SNR   Signal-to-noise ratio

SOA   Service-oriented architecture

SOAP   Simple Object Access Protocol

SONET   Synchronous optical network

SRD   Short-range devices

SS   Secondary substation

SSL   Secure sockets layer

STM   Synchronous transport module

STP   Spanning Tree Protocol

STS   Synchronous transport signal

T&D   Transmission and distribution

TACACS+   Terminal Access Controller Access Control System Plus

TC   Technical Committee

TCO   Total cost of ownership

TCP   Transmission Control Protocol

TDM   Time division multiplexing

TDMA   Time division multiple access

TEDS   TETRA Enhanced Data Service

TETRA   Terrestrial trunked radio

TFTP   Trivial File Transfer Protocol

TIA   Telecommunications Industry Association

TOS   Type of service

TOU   Time of use

TSO   Transmission system operator

TSP   Telecommunication service provider

TV   Television

TWACS   Two-way automatic communications system

UCC   Utility control center

UDP   User Datagram Protocol

UHF   Ultrahigh frequency

UMTS   Universal mobile telecommunications system

UNB   Ultranarrowband

UTC   Utilities Telecom Council

var   Volt-ampere reactive

VC   Virtual container

VDSL   Very high-speed digital subscriber line

VHF   Very high frequency

VLAN   Virtual local area network

VLF   Very low frequency

VLL   Virtual leased line

VoIP   Voice over Internet Protocol

VPLS   Virtual Private LAN Service

VPN   Virtual private network

VPP   Variable peak pricing

VPRN   Virtual private routing network

VPWS   Virtual pseudo-wire service

VRRP   Virtual Router Redundancy Protocol

VSAT   Very small aperture terminal

VT   Virtual tributary

WAMS   Wide area measurement system

WAN   Wide area network

WCDMA   Wideband code division multiple access

WDM   Wavelength division multiplexing

WG   Working Group

WiMAX   Worldwide Interoperability for Microwave Access

WLAN   Wireless local area network

WPAN   Wireless personal area network

WPS   Wireless priority service

WRC   World Radiocommunication Conference

xDSL   x Digital subscriber line (covering various types of DSL)

XLPE   Cross-linked polyethylene

XML   Extensible Markup Language

About the Authors

Alberto Sendin received an M.Sc. in telecommunication in 1996 and a Ph.D. in 2013 from the University of the Basque Country, Spain. He received an M.A. in 2001 in management for business competitiveness from the same university. Since 1998, he has been working for Iberdrola in charge of the Telecommunication Projects Department, transforming its telecommunication network. Dr. Sendin is currently a lecturer at the University of Deusto, Bilbao, Spain, where he has taught courses in telecommunication engineering since 1999. His publications (8 books, 3 book chapters, and tens of technical articles) mainly focus on radio telecommunication systems, telecommunication systems infrastructure, and power line communications (PLC).

Miguel A. Sanchez-Fornie received an M.Sc. in electrical engineering from the University of Comillas/ICAI in 1974. Mr. Sanchez-Fornie has more than 40 years of experience in the utility sector and is the Director of Global Smart Grids at Iberdrola. He is a member of the Utility Telecomms Council Board of Directors and the president of its European division. He is a member of the Advisory Committee of the European technological platform on Smart Grids, a member of the Advisory Committee of the Smart Grids Task Force (DG Energy), and a cochairman of the EG1 within the European Network Information Security (NIS) platform. He is a member of the Advisory Committees of the Massachusetts Institute of Technology Future of the Electric Grid Study and the Utility of the Future project and he chairs the Management Board of the PRIME Alliance. He is also a postgraduate university professor at the University of Comillas, Madrid, Spain.

Iñigo Berganza received an M.Sc. in telecommunication in 2000 from the University of the Basque Country, Spain. He has been working for Iberdrola since 2001 in Spain and the United States. Mr. Berganza has been involved in many activities related to the Smart Grid PLC, including the IEEE, ITU, IEC,

ETSI, and CENELEC and EU-funded projects such as OPERA, OPENmeter, and ADDRESS. He has authored several papers, articles, and book chapters on different aspects of telecommunications in the utility industry. He currently chairs the Technical Working Group of the PRIME Alliance.

Javier Simon received an M.Sc. in telecommunication in 2000 from the Technical University of Madrid, Spain. He received an M.A. in 2008 in energy company management from the University Antonio de Nebrija. Since 2001, he has worked for Iberdrola. His expertise includes telecommunication networking and PLC technology. His background includes IT consultancy.

Iker Urrutia received an M.Sc. in telecommunication in 2002 from the University of Deusto, Spain. Since 2006, he has worked for Iberdrola. His expertise includes PLC technology and application-level protocols. His background includes IT consultancy.

Index

Recent Artech House Titles in Power Engineering

Andres Carvallo, Series Editor

Signal Processing for RF Circuit Impairment Mitigation in Wireless Communications, Xinping Huang, Zhiwen Zhu, and Henry Leung

Synergies for Sustainable Energy, Elvin Yüzügüllü

Telecommunication Networks for the Smart Grid, Alberto Sendin, Miguel A. Sanchez-Fornie, Iñigo Berganza, Javier Simon, and Iker Urrutia

For further information on these and other Artech House titles, including previously considered out-of-print books now available through our In-Print-Forever® (IPF®) program, contact:

Artech House
685 Canton Street
Norwood, MA 02062
Phone: 781-769-9750
Fax: 781-769-6334
e-mail: artech@artechhouse.com

Artech House
16 Sussex Street
London SW1V 4RW UK
Phone: +44 (0)20 7596-8750
Fax: +44 (0)20 7630-0166
e-mail: artech-uk@artechhouse.com

Find us on the World Wide Web at: www.artechhouse.com